PERIODIC TALES

'"Elements are fun" is the essential premise of Hugh Aldersey-Williams's new book and by heck he's right. An entertaining trawl through the fact, myth and anecdote surrounding chemical substances. Aldersey-Williams mourns the fact chemistry isn't really sexy any more; *Periodic Tales* is a step towards it getting its mojo back' *Metro*

'Splendid. Enjoyable, polished. A glistering effort' *Observer*

'A flashily brainy book crammed with literary references and held together by a personal quest' *Daily Telegraph*

'A chemical romp from aluminium to zinc. Imaginative and fun. Almost every page yields a nugget' *Nature*

'Rumbustious, pungent with the reek of primal matter' *Independent*

'If you want an entertaining chemistry refresher course, I'd recommend it' Lauren Laverne

'A joyous romp through the chemical elements' *Today*, BBC Radio 4

## ABOUT THE AUTHOR

Hugh Aldersey-Williams studied natural sciences at Cambridge. He is the author of several books exploring science, design and architecture, and has curated exhibitions at the Victoria and Albert Museum and the Wellcome Collection. He lives in Norfolk with his wife and son.

www.hughalderseywilliams.com

# Periodic Tales

*The Curious Lives of the Elements*

## HUGH ALDERSEY-WILLIAMS

PENGUIN BOOKS

PENGUIN BOOKS

UK | USA | Canada | Ireland | Australia
India | New Zealand | South Africa

Penguin Books is part of the Penguin Random House group of companies
whose addresses can be found at global.penguinrandomhouse.com.

Penguin
Random House
UK

First published by Viking 2011
Published in Penguin Books 2012

018

Copyright © Hugh Aldersey-Williams, 2011

The moral right of the author has been asserted

The credits on p. 414 constitute an extension of this copyright page

Typeset by Palimpsest Book Production Limited, Falkirk, Stirlingshire
Printed in Great Britain by Clays Ltd, St Ives plc

A CIP catalogue record for this book is available from the British Library

ISBN: 978-0-141-04145-2

www.greenpenguin.co.uk

MIX
Paper from
responsible sources
FSC® C018179

Penguin Random House is committed to a
sustainable future for our business, our readers
and our planet. This book is made from Forest
Stewardship Council® certified paper.

To my parents
Mary Redfield Aldersey-Williams
(23 June 1930–16 May 2004)
Arthur Grosvenor Aldersey-Williams
(6 June 1929–23 December 2008)
with love and gratitude

# Contents

# List of Illustrations

All uncredited photographs are by the author.

# Acknowledgements

It must have been Andrea Sella who provided the spark that lit the fuse for this book a few years ago when he drew my attention to the curious fact that euro bank notes rely upon the element europium for their security markings. The fuse was laid long ago, however, at a time when it was hardly considered decent to explore connections between the sciences and the arts. I thank my teachers, and especially Mike Morelle and Andrew Szydlo, for encouraging the transgression that has led to this present explosion. My brother John sharpened memories of these school times.

Great thanks go to my literary agent Antony Topping at Greene & Heaton, who saw that there was a different book to be written about the elements and believed that I could write it. I am immensely grateful to Venetia Butterfield at Viking Penguin for commissioning such a self-indulgent project, and to her colleagues who pitched in with their own examples of the elements in literature, and to Sara Granger at Penguin and Andrew Cochrane at Clays, the printer of this book, who even looked into the origins of new-book smell for me. Grant Gibson, the editor of *Crafts* magazine, commissioned an article that enabled me to rehearse some of the themes I explore here. My editor Will Hammond introduced me – too late, obviously – to the term 'inkhorn', and then took the time to see that I didn't come across as one. My copy editor David Watson skilfully spared me other blushes.

I would also like to thank those writers, artists, craftspeople, curators, scientists, historians of science and others who shared some aspect of my fascination with the elements: Santiago Alvarez, Marité Amrani, Paola Antonelli, Peter Armbruster, Ken

Arnold and James Peto and Lisa Jamieson at the Wellcome Collection, London, Peter Atkins, Fiona Banner, Paola Barbarino, Fiona Barclay, Geoffrey Batchen, Bernadette Bensaude-Vincent, Jim Bettle, Michael Bierut, Lauren Bloemsma of the Telluride Historical Museum, Hasok Chang, David Clarke, Ole Corneliussen and Yanko Tihov and those behind the counter at Cornelissen's artists' supplier, Amelia Courtauld, Malcolm Crowe, Alwyn Davies, Igor Dmitriev, John Donaldson, Darby Dyar, who described the spectroscopic inspection of the surface of Mars, Matthew Eagles and Simon Cornwell, enthusiasts for sodium street lamps, Michelle Elligott, Richard Emmanuel-Eastes, Martha Fleming, Hjalmar Fors, Katie George, Irene Gil Catalina, Victoria Glendinning, Lisha Glinsman, who discovered that it was lead that gave Rodin's *Thinker* his bottom, Antony Gormley, Clare Grafik at the Photographers' Gallery, Karl Grandin and Anne de Malleray at the Royal Swedish Academy of Sciences, Carol Grissom, Domingo Gutierrez, the mayor of Boron, California, Eva Charlotte and Lutz Haber, Hans de Heij, Julian Henderson, Richard Herrington, Kate Hodgson, Erika Ingham, Frank James at the Royal Institution of Great Britain, David Jollie and Keith White at Johnson Matthey, Graeme Jones, John Jost of the International Union of Pure and Applied Chemistry, Chris Knight, Susanne Kuechler, Peter Lachmann, Charles Lambert, Ron Lancaster, Petra Lange-Berndt, Anders Lundgren, Clare Maddison of Contemporary Applied Arts, Jim Marshall, Marcos Martinón-Torres, Pauline Meakins, Andrew Meharg, Andries Meijerink, Anne Mellows at the Museum of Brands, London, Jacqueline Mina, Mark Miodownik, Zoe Laughlin and Martin Conreen, keepers of the materials library at King's College London, John Morgan, Andrew Motion, Tessa Murdoch, Thierry Nectoux, Margaret Newman at the Royal Naval Museum, who told me about the various ships *Sulphur*, William Newman, Pati Núñez, Peter

Oakley, Yuri Oganessian, Cornelia Parker, Tim Parks, Simon Patterson, David Poston, Pekka Pyykko, Renny Ramakers, Jeffrey Riegel, Charlotte Schepke, Ann Marie Shillito, the late Sir Reresby Sitwell, Hans Stofer, Freek Suijver, Camilla Sundvall, Grainne Sweeney and Alex Evans at the National Glass Centre, Sunderland, Peter Tandy, Nicolas Thomas, Jan Trofast, Janet Vertesi, Luba Vikhanski, Peter Waldron and Paul Robinson and the staff of Winsor & Newton, Jo Warburton, Martijn Werts, Gull-Britt Wesslund, Max Whitby, Gavin Whittaker, David Wright.

My thanks are also due to the staff of the Cambridge University Library, the very design of which does so much to facilitate the kind of boundary-crossing exploration I have attempted. John Emsley's magisterial *Nature's Building Blocks* was never far from my side, and a number of websites, notably those maintained by Peter van der Krogt and Theodore Gray, furnished me with additional background.

Above all, I thank my wife Moira and son Sam, who offered encouragement and showed the greatest enthusiasm for this odd and wonderful project.

Hugh Aldersey-Williams
Norfolk, June 2010

# Prologue

Like the alphabet or the zodiac, the periodic table of the elements is one of those graphic images that seem to root themselves for ever in our memories. The one I remember is from school, hung on the wall behind the teacher's desk like an altar screen, its glossy yellowing paper testament to years of chemical attack. It's an image I haven't been able to shake off, despite scarcely venturing into a laboratory for years. Now I have it on my own wall.

Or at least a version of it. The familiar stepped skyline is there, and the neatly stacked boxes, one for each element. Each box contains the symbol and atomic number appropriate to the element in that position. However, all is not quite as it should be in this table. For where the name of each element should appear, there is another name entirely, one that is nothing to do with the world of science. The symbol O represents not the element oxygen but the god Orpheus; Br is not bromine but the artist Bronzino. Many of the other spaces are taken, for some reason, by figures from 1950s cinema.

This periodic table is a lithograph by the British artist Simon Patterson. Patterson is fascinated by the diagrams that are the means by which we organize our world. His way of working is to recognize the importance of the thing as an emblem of order but then to play havoc with its contents. His best-known work is a London Underground map with the stations along each line renamed after saints and explorers and football players. Strange things happen at the intersections.

It is no surprise that he should wish to play the same game

with the periodic table. He has grim memories of how it was taught by rote at his school. 'It was convenient to teach it that way, but I could never remember it,' Simon tells me. Yet he remembered the *idea* of it. Ten years after leaving school, he produced a series of variations on the table in which the symbol for each element kicks off a false association. Cr is not chromium but Julie Christie, Cu not copper but Tony Curtis; and then even this cryptic system is sabotaged: Ag, the symbol for silver, is not Jenny Agutter, say, or Agatha Christie, but of course Phil Silvers. There are teasing moments of apparent logic in this new tabulation: the sequential elements beryllium and boron (symbols Be and B) are the Bergmans, Ingrid and Ingmar respectively. The acting brothers Rex and Rhodes Reason appear adjacent to one another, co-opting the symbols for rhenium (Re) and osmium (Os). Kim Novak (Na; sodium) and Grace Kelly (K; potassium) share the same column in the table – both were Hitchcock leading ladies. But in general there is no system, only the connections you make for yourself: I was tickled to see, for example, that Po, the symbol for polonium, the radioactive element discovered by Marie Curie and named by her for her native Poland, denotes instead the Polish director Roman Polanski.

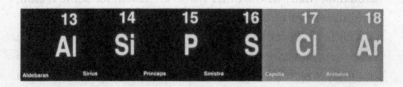

I now love the ludic irreverence of this work, but my school-age self would have been quite scornful of such nonsense. While Simon was dreaming up wild new connections, I was merely absorbing the information I was meant to absorb. The elements, I understood, were the universal and fundamental ingredients of all matter. There was nothing that was not made out of elements. But the table into which the Russian chemist Dmitrii Mendeleev had sorted them was even more than the sum of these remarkable parts. It made sense of the riotous *variety* of the elements, placing them sequentially in rows by atomic number (that is to say, the number of protons in the nuclei of their atoms) in such a way that their chemical relatedness suddenly leapt out (this relatedness is *periodic*, as revealed in the alignment of the columns). Mendeleev's table seemed to have a life of its own. For me, it stood as one of the great and unquestionable systems of the world. It explained so much, it seemed so natural, that it must always have been there; it couldn't possibly be the recent invention of modern science (although it was less than a century old when I first saw it). I acknowledged its power as an icon, yet I too began to wonder in my own tentative way what it really meant. The table seemed in some funny way to belittle its own contents. With its relentless logic of sequence and similarity, it made the elements themselves, in their messy materiality, almost superfluous.

Indeed, my classroom periodic table provided no picture of what each element looked like. The realization that these ciphers had real substance struck me only at the vast illuminated table of the chemical elements they used to keep at the Science Museum in London. This table had actual specimens. In each rectangle of the already familiar grid squatted a little glass bubble beneath which a sample of the relevant element glimmered or brooded. There was no knowing whether they were all the real thing, but I noted that the curators had omitted to include many of the rare and radioactive elements, so it seemed safe to assume that the rest were authentic. Here it was vividly clear what we had been told at school: that the gaseous elements were mostly to be found in the top rows of the table; that the metals occupied the centre and left, with the heavier ones in the lower rows – they were mostly grey, although one column, containing copper, silver and gold, provided a streak of colour; that the non-metals, more variegated in colour and texture, lay over in the top right corner.

With that, I had to start my own collection. It would not be easy. Few of the elements are found in their pure state in nature. Usually, they are chemically locked up in minerals and ores. So instead, I began to cast about the house, taking advantage of the centuries during which man has extracted them from these ores and pressed them into service. I broke open dead light bulbs and surgically snipped free the tungsten filaments, placing the wriggling wires into a little glass vial. Aluminium came from the kitchen in the form of foil, copper from the garage as electrical wire. A foreign coin that I'd heard was made of nickel – though not an American nickel, which I knew was mostly copper – I cut up into coarse chunks. It was worth more to me like this. It made it more, well, elemental. I discovered that my father had some gold leaf kept from his youth, when he used it for decorative lettering. I removed some of it from the drawer where it had lain in darkness for thirty years and allowed it to shine once more.

This was a definite improvement on the Science Museum. I could not only see my specimens close up, but feel whether they were warm or cold to the touch and heft them in my hand – a bright little ingot of tin, which I had cast in a small ceramic bath from a melted roll of solder, was astonishingly heavy. I could make them ring or rattle against the glass and appreciate their characteristic timbres. Sulphur had a primrose colour with a slight sparkle, and could be poured and spooned like caster sugar. For me, its beauty was in no way tainted by its slightly pungent odour. I have reminded myself of this smell just now, with a tin of sulphur bought from a garden shop, where it is sold to fumigate greenhouses. The dry, woody aroma is on my fingers as I type, to me not hellish as the Bible teaches, but evocative simply of childhood experimental enquiry.

Other elements needed more work. Zinc and carbon came from batteries – zinc from the casing, which serves as one electrode, and carbon from the rod of graphite inside it that provides the other. So did mercury. More expensive, mercury batteries were used to run various electronic gadgets. By the time they had run down, the mercuric oxide that powered them had been reduced to metallic mercury. I chopped off the ends of the batteries with a hacksaw and scooped out the sludge into a flask. By heating the flask, I was able to distil off the metal, watching as tiny glistening droplets condensed from the thick toxic fumes and then merged into a single hyperactive silvery bead. (The experiment would be banned now for health reasons, as are these batteries.)

A few of the elements you could still buy, in those innocent days, at a dispensing chemist's. I got my iodine in this way. Others came from a small chemicals supplier in Tottenham long since driven out of business by restrictions on the sale of what were of course the raw materials for bombs and poisons – as well as everything else. Although my parents were happy enough to indulge my obsession by driving me there, these trips along the farther reaches of the Seven Sisters Road to the shabby counter beneath the thundering railway arches, with its aromas as promising as any spice market, always had a clandestine feel about them.

I made good progress with my table. I had drawn the grid out on a backboard of plywood and hung it on the bedroom wall. As I got it, I dropped each new sample into a uniform vial and clipped it into position on the grid. The pure elements themselves were often chemically rather useless. I saw that. The useful chemicals – the ones that reacted or exploded or made beautiful colours — were mostly the chemical combinations of elements known as compounds, and these I kept in a cupboard in the bathroom where I did my experiments. The elements were a collector's obsession. They had a beginning and a compelling sequence. They seemed also to have an end. (Little did I know then of the ferocious cold war between American and Soviet scientists, who were striving to add to the 103 I had fixed in my head by synthesizing new ones.) As a collector, my aim, however unattainable it was destined to be, was of course to complete the set. But it was far more than collecting for collecting's sake. Here I was assembling the very building blocks of the world, of the universe. My collection had none of the artifice of stamps or football cards, where the rules of the game are set arbitrarily by other collectors or, worse still, by the companies producing the items in the first place. This was fundamental. The elements were for ever. They had come into being soon after the Big Bang, and would be here long after humankind has perished, after all life on earth, even after the planet itself has been consumed by its own ballooning red sun.

This was the system of the world that I chose — a system as complete as any other on offer. History, geography, the laws of physics, literature: each was all-embracing according to its lights. Everything that happens happens in history, has its place in geography, is reducible solely to the interaction of energy and matter. But it is also materially constituted of the elements, no more and no less: the Great Rift Valley, the Field of the Cloth of Gold, Newton's prism, the *Mona Lisa*; all impossible without the elements.

At school around this time, we were reading *The Merchant of Venice*. I was Bassanio for one forty-minute session — not a bad role, though I loathed reading out loud. We came at length to the scene when it is Bassanio's turn to select the one of three caskets that contains Portia's likeness in order that he might win her hand in marriage. The unlucky boy who was Portia prattled on while I waited in dread for my entrance. 'Let me choose / For as I am, I live upon the rack,' I intoned with no feeling whatever. Then I was having to choose between the imaginary caskets. I am sure nobody could have gleaned anything of my character's reasoning from my featureless voice as I rejected first the 'gaudy gold' and then the silver, 'thou pale and common drudge / 'Tween man and man', before plumping for 'meagre lead'. But somewhere inside my head something clicked. Three of the elements! Was Shakespeare a chemist? (Later, I found that T. S. Eliot was a chemist too, a spectroscopist in fact: in *The Waste Land*, he presents a vivid image as a nail-studded ship's timber 'fed with copper / Burned green and orange' – green from the copper, orange from the sodium in the sea salt.)

Dimly, I began to perceive that the elements told cultural stories. Gold *meant* something. Silver meant something else, lead something else again. Moreover, these meanings arose essentially from chemistry. Gold is precious because it's rare, but it's also considered gaudy because it is one of the few elements that naturally occurs in its elemental state, uncombined with others, glittering boldly rather than disguised as an ore. Was there, I wondered, such a mythology for all the elements?

Their very names often spoke of history. Elements discovered during the Enlightenment had names based on Classical mythology – titanium, niobium, palladium, uranium, and so on. Those found during the nineteenth century, on the other hand, tended to reflect the fact that they – or their discoverers – were sons and daughters of some particular soil. The German chemist Clemens Winkler isolated germanium. The Swede Lars Nilson named his discovery scandium. Marie and Pierre Curie found polonium and named it – not without encountering some resistance – after Marie's fondly remembered homeland. A little later, the scientific spirit grew more communitarian. Europium was named in 1901 – and towards the end of that new century some humorous bureaucrat in one of Europe's banks would decree that compounds of this element should be used for the luminescent dyes that are incorporated into euro bank notes for the easier detection of counterfeits. Who would have thought it? Even obscure europium has its cultural day.

So the elements inhabit our culture. We should not really be surprised at this: they are the ingredients of every thing, after all. But we should be surprised at how seldom we notice this fact. This missed connection is partly the chemists' fault for presuming to study and teach their subject in lofty isolation from the world. But the humanities are also to blame: I was astonished to find, for instance, that a biographer of Matisse could complete her work without saying what pigments the artist used. Perhaps this makes me unusual, but then again I'm sure Matisse cannot have been indifferent to the matter.

The elements do not simply occupy fixed spaces in our culture as they do in the periodic table. They rise and fall on the tide of cultural whim. John Masefield's famous poem 'Cargoes' lists eighteen commodities in its three short verses portraying three eras of global trade and plunder, eleven of them either elements in their pure state or materials which derive their value from the particular nature of one element ingredient, from the quinquereme of Nineveh with its calcareous white ivory to the dirty British coaster with its load of 'Tyne coal, / Road-rails, pig-lead, / Firewood, iron-ware, and cheap tin trays'.

From the moment of its discovery, each element embarks upon a journey into our culture. It may eventually come to be visible everywhere, like iron or the carbon in coal. It may loom large economically or politically while remaining largely unseen, like silicon or plutonium. Or it may, like europium, provide a grace note only appreciated by those in the know.

When I wrote my school essays ('Why does Bassanio choose the lead casket?') it was with an Osmiroid pen, a brand name inspired by the osmium and iridium that its manufacturer used to harden the nibs.

During its gradual assimilation, we come to understand the element better. The experience of those who mine it, smelt it, shape it and trade it gives it meaning. It is through these muscular processes that an element's weight is felt and its resistance is gauged, so that Shakespeare can then refer to gold and silver and lead in the ways that he does knowing that his audience will understand him.

It is not only the ancient elements that are culturally involved. Contemporary artists and writers have used relatively new-found elements such as chromium and neon to send particular signals just as Shakespeare used the elements known in his day. These elements, which fifty years ago signified the innocent glamour of the consumer society, now seem to us tawdry and full of empty promise. The place once occupied by 'chrome' is now perhaps taken by a newer element, 'titanium', which brands fashionable clothing and computer equipment. In such cases, the element's meaning detaches itself almost completely from the element itself: how many more platinum blondes and platinum credit cards (neither incorporating any platinum) there must be than platinum rings. Even some highly recherché elements undergo this shift. 'Radium' was once popular, sometimes in substance, sometimes in name alone, for all manner of health remedies. There are no longer Osmiroid pens, but there is an Iridium telephone company.

If I were to reassemble my periodic table now, I would still want to include a specimen of each element, but I would also want to trace its cultural journey. I feel that the elements leave great streaks of colour across the canvas of our civilization. The black of charcoal and coal, the white of calcium in chalk and marble and pearl, the intense blue of cobalt in glass and china slash boldly through place and time, geography and history. *Periodic Tales* is the start of that collection.

It is therefore a book of stories: stories of discovery and of discoverers; stories of rituals and values; stories of exploitation and celebration; stories of superstition as well as science. It is not a chemistry book – it contains as much history, biography and mythology as chemistry, and generous helpings of economics, geography, geology, astronomy and religion besides. I have purposely avoided discussing the elements in their periodic table sequence or giving a systematic description of their properties and uses. Other books do this well. I believe that the periodic table has become an icon too powerful for its own good. The ordered grid of squares with its raggedy edges, the strange names and cryptic symbols, the way the elements follow a sequence as fixed yet as apparently arbitrarily as the letters of the alphabet: all these things are strangely compelling. They provide limitless raw material for television quizzes: what element lies directly south-east of zinc?* Who cares? Even chemists do not use the table in this way.

The elements provide the real interest. The periodic table that I once thought of as unquestionable I know now does not really *exist*. A few chemists might deny it, but it *is* only a construct, a mnemonic that arrays the elements in a particularly clever way so as to reveal certain commonalities among them. Yet there's no actual law against arranging the elements by different rules. In his famous song 'The Elements', the American satirist Tom Lehrer reordered them purely for the sake of rhyme and scansion, to fit the tune of Arthur Sullivan's patter song, 'I am the very model of a modern major-general', from *The Pirates of Penzance*.

I wish to discover the *cultural* themes that group the elements

* The answer is indium.

anew, to draw up the periodic table as if sorted by an anthropologist. To this end, I have chosen five major headings: power, fire, craft, beauty and earth.

As Masefield's poem shows, imperial might has always depended on possession of the elements. The Roman Empire was built on bronze, the Spanish on gold, the British on iron and coal. The balance of the twentieth-century superpowers was maintained by a nuclear arsenal based on uranium and the plutonium made from it. In 'Power', I consider some of these elements that have been amassed as riches and, ultimately, used as a means of exerting control.

In 'Fire', I discuss those elements whose burning light or corrosive action are the key to our understanding of them. We may remember from school that sodium, for example, is an element that entertainingly explodes on contact with water, but we *know* it above all as the ubiquitous mango-yellow colour of our street lamps – a very particular light that many writers have seized upon as the index of a general urban malaise.

In the end, any cultural meaning that an element acquires derives from its fundamental properties. This is seen most clearly in the case of those elements that craftspeople have chosen as their raw material. It is the centuries or millennia of hammering and drawing, casting and polishing that have given many of the metallic elements their meaning. 'Craft' explains why we regard lead as grave, tin as cheap and silver as radiant with virginal innocence.

Humankind has manipulated the elements not only for their utility, but also for the sheer joy of the look of them. 'Beauty' shows how the compounds of many of the elements, and the light of others, colour our world. Finally, in 'Earth', I travel to Sweden to discover how particular places have marked many of the elements, and how those places are marked in turn by the fluke of finding an element there.

My own journey has led me to mines and artists' studios, to factories and cathedrals, into the woods and down to the sea. I have recreated early experiments in order to make a few of the elements for myself. I have been pleased to find the elements in abundance in fiction, too, where Jean-Paul Sartre sees fit to remark upon the constancy of the melting point of lead (335 degrees centigrade, he says) and Vladimir Nabokov finds mandalic significance in the carbon atom 'with its four valences'. Wandering through Shoreditch in London on my way to see Cornelia Parker, an artist who has made it her business to remind us of the cultural significance of many of the elements, I was captivated by a sculpture in a shop window by some other artists of a nuclear power station wittily cast like lime jelly in glowing uranium glass. It was clear. The elements do not belong in a laboratory; they are the property of us all. *Periodic Tales* is a record of the journey with the elements that I was never encouraged to take when I was a chemist. Come along: there will be fireworks.

# Part One: Power

## *El Dorado*

In 2008, the British Museum commissioned a life-size sculpture of the model Kate Moss. The artwork, called *Siren*, is made entirely of gold and is said to be the largest gold sculpture created since the days of ancient Egypt, though it's impossible to check whether this is true. *Siren* was placed on show in the museum's Nereid Gallery near a statue of a bathing Aphrodite. My immediate impression on seeing Kate Moss's otherwise familiar image is how tiny she looks, accentuated by the fact that she is knotted in a particularly uncomfortable-looking yoga position, though this may be an optical illusion – we are unused, after all, to seeing so much of the shining metal at once. The gold, I am disappointed to find, is not polished to a high gloss but has a steely brushed finish, which elicits a high sparkle from the grains in the textured surface, not the burnished glow I had expected to see. There are signs of pitting in the casting, which a different goldsmith might have taken care of. The unique qualities of the metal that have made it precious to cultures since antiquity seem poorly served. Only the face is perfectly smooth, and is immediately reminiscent of the funerary mask of Tutenkhamun. The lifeless staring visage has the disturbing effect, entirely unexpected given the high public profile of its subject, of plucking the spectator out of time: this is no longer a rendering of the twenty-first-century celebrity, but a depersonalized, detemporalized figure whose sharp nose and pouting lips belong less to a living person than to a death mask or votive figure.

The price put on the statue was £1.5 million. It was the whim of the artist, Marc Quinn, that the work be fabricated from gold of equal mass to the model's fifty-kilogramme body, so that in addition to appearing life-size, it could be said to represent her weight in gold, perhaps raising in the mind of the astute onlooker thoughts of ransom and slavery. In solid gold, I calculate, Kate would be shrunk to the size of a garden ornament. Quinn's piece must therefore be hollow, which may also be an artistic comment of some kind. Although gold is the only declared material from which the work is made, I figure there must be some sort of armature to support the weight of the soft

metal, which would otherwise slump out of shape. Afterwards, I look up the price of gold. Although *Siren* went on display during a period of global financial upheaval, when the price of gold had doubled, it was still only £15,000 a kilo, giving the artwork a scrap value, as it were, of £750,000. Presumably the rest of the £1.5 million is to cover labour.

I watch as people queue to take photographs of the golden Moss, either simply snapping her image or sometimes placing their partner in the shot next to her, making who knows what sort of comparison. I am curious to know what has drawn them to the sculpture. Which is more powerful: the cult of celebrity or the cult of gold? What is really the siren here? It is mainly men who have come to worship this modern Aphrodite. A few purport to admire the sculptural qualities of the work. Some are indeed drawn by the power of celebrity, but are fans of Quinn more than Moss. I ask the girlfriend of one temporarily distracted Polish man what she thinks of it. 'It is beautiful,' she concedes, as if to say otherwise would be unacceptable, 'but it doesn't belong here.' Another woman photographing it with her phone is briskly dismissive: 'I need some gold for my mobile – it's wallpaper.'

More than any of the ancient elements, gold has been judged to possess a timeless allure. None of the elements discovered by modern science has challenged this supremacy. But what, if anything, is truly special about this metal?

Gold is characteristically yellow. In a flower, one might find this yellow attractive or not – beauty is a matter of taste, after all. But in gold, apparently, the unique combination of this colour with the lustre of metal leaves us no other option than to be drawn to it. Even the sociologist Thorstein Veblen, whom one might expect to maintain some professional caution in the matter, falls for the stuff. In a chapter on the 'pecuniary canons of taste' in his classic text *The Theory of the Leisure Class* (1899), he writes that gold has a 'high degree of sensuous beauty' as if that were an objective fact, and never mind the eye of the beholder.

There is then the fact that this colour and lustre endure, because gold resists corrosion by the air, by water and indeed by almost all chemical reagents. Pliny the Elder thinks it is this unique quality of endurance, and specifically *not* its colour, that explains our love of gold: 'it is the only metal that loses nothing by contact with fire', he observes. It is this endurance that gives gold its association with immortality, and so with royal lines and divinity. The Buddha is gilded as an indication of enlightenment and perfection, and the metal's incorruptibility inspires a torrent of other ideals: the golden section, the golden mean, the golden rule.

Gold is special also because of its great density, its malleability and ductility – it can be beaten as thin as hair and 'long enough to encircle a whole village', as one West African proverb puts it. It is surely the case that gold's heaviness, in particular, signifies value in the way that dense materials often do, regardless of their actual composition, because their relative weight transmits a sense of sheer quantity. Gold's resistance to chemical attack – in other words its ability to retain its pure state – signifies value too, because we naturally place value in things that endure. It is these economically important secondary attributes of the element that give Veblen cause to comment on it at all. And it is this muddled equation between beauty and value that lies at the heart of our understanding of gold.

Though gold was known to the ancients, being the only metal typically found in the elemental state, it was too soft for making weapons and was perhaps not much used at first, even for ornamental purposes. Even where it is relatively abundant, such as in parts of Australia and New Zealand, aboriginal peoples have often ignored it. In Europe, Africa and Asia, however, the metal was generally highly prized and was soon taken up for jewellery and then for coin. The first coins were stamped out of electrum, a natural alloy of gold and silver, in Lydia in the seventh century BCE. Around 550 BCE King Croesus minted purer silver and gold coins, and from then on the yellow metal was man's chosen element for the expression of great wealth. Backed by state authority, Croesus' coinage boosted trade and banking. For gold to hold its greater value as coin against native electrum it had to be pure, and its purity had to be ascertainable by assay. With this, gold became subject to comparative testing and valuation as well as absolute worship.

Six hundred years later, Pliny is scathing about the corrupting effect of gold, which he wished 'could be completely banished from life'. He damns equally those who wear it and those who trade with it: 'The first person to put gold on his fingers committed the worst crime against human life.' 'The second crime against mankind was committed by the person who first struck a gold denarius.' The difficulty lies not with the material itself, but with man's transforming hands upon it. Natural gold may contain the light of the sun, but minted gold becomes a 'symbol of perversion and the exaltation of unclean desire'. Sir Thomas More confirms this moral distinction in his *Utopia*, reserving its gold not for finery but for making chamberpots.

Harder heads have always understood that gold is the key to power. Had not the Pharaohs reigned for 3,000 years relying on their gold to contain the more ingenious Sumerians and Babylonians? Had not the Romans been driven to conquest by their envy of the gold possessed by the Gauls, the Carthaginians and the Greeks?

Such is the monetary value of gold that natural deposits tend to acquire an aura so dazzling that they soon become detached from any real geography. Ophir was the biblical source of Solomon's gold. It is the port, probably in southern Arabia, from which sails the gold-laden quinquereme of Nineveh in John Masefield's 'Cargoes'. Strabo's *Geographica* mentions gold mining on the African bank of the Red Sea, presumably one source of the Egyptians' gold. But as the means expand so do the imaginative horizons. By the time of the Portuguese navigator Vasco da Gama, the best advice was that Ophir lay in southern Africa, roughly where Zimbabwe is today, or perhaps in the Philippines. Columbus thought Ophir was to be found on the island of Hispaniola. With the Spanish expeditions to the New World came new stories of fabulous gold and a new myth of El Dorado. El Dorado, literally 'the golden man', was said to be a tribal priest who was covered in gold for the performance of a sacred ritual, but in the imagination of Western explorers it became another unmapped place of riches, a new Ophir.

In March 1519, Hernando Cortés set out on such an expedition, sailing from Cuba with eleven ships and a force of 600 men to claim the mainland of Mexico and its treasure for the Spanish crown. After various skirmishes, Cortés reached the Aztec capital Tenochtitlan, where he and his men were ceremonially received by the emperor Montezuma II and showered with gifts of gold. By means of a subterfuge during the hospitality the Spanish managed to take Montezuma prisoner; before long the Aztec empire had fallen, and Spain was in control of most of Mexico. For all their victory, however, Cortés's men found little gold besides the presents they had been given by their hosts. It was left to later settlers to develop the Mexican silver mines that would bankroll the Spanish empire.

Thirteen years later, Francisco Pizarro, after lengthy preparations including a voyage of reconnaissance down the Pacific coast to the northern fringe of the Inca empire and another back to Spain to obtain funding, set forth to Peru in search of Inca treasure. Once again betraying the hospitality they were shown (Pizarro had been coached by Cortés back in Spain), the conquistadores launched a surprise attack and captured the Inca ruler Atáhualpa. As before, their plan was to control the territory by holding him as vassal ruler. But Atahualpa had another idea, a ransom calculated to appeal to the Spaniards: he bargained his freedom in exchange for a room, some six metres by five metres, that would be filled once with gold and twice with silver as high as a man could reach. This 'ransom room' still survives in Cajamarca, Peru. It is clear it cannot have been literally filled. Nevertheless, the Spaniards melted down some eleven tonnes of handsomely crafted gold artefacts for transport as bullion back to Spain. As the ships set sail, they reneged on the deal and put Atahualpa to death.

These were great windfalls. But where was El Dorado? The search went on. Pizarro's half-brother Gonzalo set off inland from Quito in Ecuador in 1541, but found no city of gold, only a route to the Atlantic Ocean via the River Amazon. Other Spanish adventurers heard stories of the Muisca people of Colombia, who threw gold offerings into a mountaintop lake in order to appease the golden god supposed to live at the bottom of it. When they arrived, they rudely set about trying to drain the lake, but in 400 years only a few pieces of gold have been dredged up.

In 1596, Walter Ralegh sailed to Venezuela, coming away with little gold but his belief in El Dorado nevertheless intact. Accounts of these voyages gave Voltaire plenty of material with which to lampoon the rapacity of the Europeans in his picaresque novella of 1759, *Candide*. The naive hero Candide is expelled from his vapid and paradisiacal life in Westphalia to travel the world and witness its hardships, from the Thirty Years' War to the Lisbon earthquake. He finds Eldorado with no trouble and, after being royally entertained, is sent on his way with gifts of fifty sheep laden with gold and jewels. At first, Candide and his companions are buoyed up by the vision of themselves as the 'possessors of more treasures than Asia, Europe and Africa could gather', but as they travel on, the sheep fall by the wayside in ones and twos, bogged down in swamps, or fallen from preci-pices, forcing Candide to acknowledge 'how the riches of the world are perishable'.

Between 1520 and 1660, Spain imported 200 tonnes of gold, never finding it in one convenient hoard, but by expanding its mining activities throughout its territories in the New World. El Dorado was never a place; always an idea.

What these recurrent episodes have in common, apart from European greed and treachery, is the presumption that all parties are agreed that gold is the most valuable substance known to man. This was nothing like the case. The Aztecs and the Incas and other New World indigenous peoples made golden offerings to the gods but did not use the metal for money, so it had little tradeable value, and in some cases other metals were more desirable even for religious purposes.

The Taíno inhabitants of Hispaniola, Cuba and Puerto Rico, for example, assigned distinct roles to gold and silver, and also to a range of coloured alloys. These natives, treated as slaves by Columbus and his followers, found a friend in Bartolomé de Las Casas, the first Christian priest to be ordained in the New World. Las Casas was the author of a history of the Indies, a founder of utopian communities and a believer in liberation theology who thought Cortés a vulgar adventurer. He observed the Taíno customs and found that they did not prize gold for its weight or colour, or regard it as self-evidently valuable as the Spaniards did. The Taíno placed more importance on *guanín*, an alloy of copper, silver and gold. What pleased them about it was its reddish-purplish colour and most of all its peculiar smell, probably arising from a reaction between the copper and the grease of human fingering. Pure gold, by contrast, was yellow-white and odourless and unappealing. Both gold and *guanín* were associated with power, authority and the supernatural world, but *guanín* carried the greater symbolic charge. Unlike gold, which was found native, *guanín* had to be smelted. This made the alloy still more precious, especially since the technology was not available on Hispaniola and had to be imported from Colombia, which made it seem as if it came from another world. Gold could be dredged up from river-beds, but it seemed as if *guanín* could only be made in heaven.

Brass, an Old World alloy entirely unknown to pre-Columbian societies, had the same attractive qualities as *guanín*. Brought by the Spanish, it too was seen as coming from the remote heavens and given a local name likening its brightness to the sunny sky. How much did gold appreciate in value with each nautical mile on its eastward journey to Spain? And how much did humble brass gain as it sailed west? The image of Spanish ships ferrying the two yellow metals each way across the Atlantic with no other point than to feed two mutually uncomprehending societies' taste in luxury is one to bring an ironic smile to the lips of any Veblen or Voltaire.

I feel it is time I got my hands on some gold, and arrange to meet Richard Herrington, an economic mineralogist at the Natural History Museum in London and an authority on the stuff. The floor of his office is strewn with variegated rocks, red ochre, glittering white, metallic black, each snug in its own box. I have to pick my way carefully just to take a seat. Herrington himself wears a lumberjack shirt as if he has just come in from the mountainside. 'I love gold,' he says simply. 'I love finding it in the rock.' He hands me a paperweight-sized piece of quartz with a dark yellow inclusion of gold the size of a fingernail. 'Everybody understands gold. We've seen it in the credit crunch. It's an alternative and trusted commodity. Even a popular paper will quote the gold price every day.' The value of a diamond depends on its optical quality, that of a painting on everybody else's opinion of the artist. But gold is always gold, pure and simple. 'I can't see it being replaced.'

Gold became a more democratic pursuit with the gold rushes of the nineteenth century. The American president James Polk inadvertently launched the first of these when he mentioned that gold had been found at Sutters Fort, California, in his annual statement to Congress in December 1848. By the end of 1849, the non-native-American population of the state had quadrupled to 115,000. In Australia not long after, the British crown attempted to assert its medieval prerogative over gold mines, but the gold rush was so frenzied, and the administration so inept, that it could not be enforced. Repeated over and over again in North America, Australia and elsewhere up until the early years of the twentieth century, the rush for gold, and the consequent increase in gold production, led economists, unable to see the metal as anything other than coin currency, to fear a wholesale collapse in the value of money itself.

One of the early American prospectors was Samuel Clemens, who only became the writer we now know as Mark Twain when he failed in his quest for gold. Clemens went west in 1861 to the Nevada Territory, where his brother was the governor. He tried his luck at several seams and wrote about his experience in his memoir, *Roughing It*. The memoir is peppered with the grandiloquent names given to the modest ledges and seams he acquired as stakes, but it also betrays Twain's sheer distaste for the work, repeatedly blasting and sifting the 'hard, rebellious quartz' to obtain the tiniest specks of colour.

Twain had every reason to feel disheartened, for he may even have finished his spell as a prospector down on the deal. Having failed to find gold, he fetched up in Virginia City, Nevada, and took a job in an ore mill separating the precious metal from the dross. One means of doing this was by amalgamation, using mercury to dissolve the gold, which could then be recovered from the amalgam by heating. Unfortunately, Twain neglected to remove the gold ring he habitually wore, which he soon observed had crumbled to pieces under attack from the mercury.

The gold may now be gone, but evidence of the rush remains in the towns that sprang up when a major deposit was found. Years ago, I visited Cripple Creek, in the high valleys of Colorado, once the site of the world's biggest gold mine. The story of the town began when a rancher, Robert Womack, found ore there in 1890. The ore was a rare mineral that contained silver and gold in the form of salts rather than as native metals. One version has it that the discovery was made when the heat from a furnace hearth caused the ground to sweat with molten gold. The prospectors came, and a year later, on the Fourth of July, a carpenter, Winfield Stratton, laid claim to Independence lode, one of the largest gold deposits ever found. In 1900, Stratton sold his mine for $10 million, while Womack drank away what little money he made. Cripple Creek eventually yielded some $300 million in gold.

I walked the length of the broad main street, a gently curving dip like the track of a pendulum. At each end, vistas opened towards snow-covered mountains with the geology naked above the tree line. The buildings that lined the street – an ice-cream parlour, a general store, a few craft shops, the boarded-up Phenix Block yet to rise again – sported a rich variety of Victorian ornamentation in brick and plaster and were overhung by elaborate wooden cornices. Many of them bore the date, the same in every case: 1896. A town that had grown from nothing in a year and where nothing had happened since. It was easy to picture the mad excitement of the rush that made these places overnight and then almost as quickly left them to rot. I noticed the offer of 'free gold ore samples' at Frego's Emporium. It seemed to confirm that the great days were over. (People still seek instant riches today even if they are no longer prepared to work for it: the town has recently sought to revive its fortunes by introducing legalized gambling.)

Mythology has often associated gold with water. The Phrygian king Midas washes away his curse of the golden touch in the River Sardis, while the story of the Golden Fleece originates from the trick of placing wool in a running stream to catch fine particles of precious metal. It is no surprise then to learn that scientists too have directed their quest beneath the waves. The Swedish chemist Svante Arrhenius, the first director of the Nobel Institute, made significant strides in many fields, including prescient speculations about the greenhouse effect in the earth's atmosphere. Much of his research was done on the electrical conductivity of solutions, during the course of which, in 1903, he arrived at an estimate of the amount of gold dissolved in the sea. His calculations put the concentration of the element at six milligrammes per tonne of seawater. At this level, the total reserve of gold in the world's oceans would be eight billion tonnes. The global annual production of gold at the time was a few hundred tonnes.

In May 1920, Arrhenius's German friend Fritz Haber travelled to Stockholm to collect the Nobel Prize that he had been awarded (for the year 1918, but postponed because of the First World War) for his discovery of a synthetic route for making ammonia from nitrogen in the atmosphere, a breakthrough that had quickly proved vital for the manufacture of both fertilizer and explosives. The two men held long discussions. A few days after Haber arrived back in Germany, the victorious Allies announced their peace terms: his country was to pay 269 billion goldmarks in reparations. He resolved to find the money using science.

Somewhere at the back of his mind must have been the legend of the Rhine gold. In the first opera of Wagner's Ring Cycle, *Das Rheingold*, the gold appears gleaming in the sunlight at the bottom of the river, guarded by three flippertigibbet Rhine maidens. The dwarfish Alberich eyes the girls, but settles for the gold and the secret they whisper to him that a ring made from it will confer upon the wearer power without limit. In common with Pliny and the great German metallurgist Agricola,

Wagner is at pains to make clear that the native metal is quite innocent in all this, and it is only objects made from it by human art that are corrupting. As George Bernard Shaw explains in *The Perfect Wagnerite*, his critique of the Ring Cycle, the Rhine maidens value the gold 'in an entirely uncommercial way, for its bodily beauty and splendor'. They sing that only man has the craft to fashion the gold into a ring, which is naturally what the jilted, venal Alberich goes and does. Over the course of the next three nights in the opera house, the ring is traded, stolen, fought over and paid as ransom, working its curse as it goes, until finally the river reclaims its own. It is perhaps significant that Wagner wrote the libretto of the cycle at the time of the first great gold rushes, while Shaw used the Klondike gold rush of 1898 to illustrate his criticisms.

The curse worked more slowly on Haber. He launched into the project by calling in samples of seawater from around the world to his Berlin laboratory. The chemical analyses confirmed Arrhenius's figures. Then, backed by a consortium of metals interests, he equipped a ship and put to sea in 1923. But on this transatlantic journey and on subsequent voyages in other oceans over the next four years, his measurements seemed to show less and less of the precious metal. He concluded despondently – and, it now appears, erroneously – that there was only a tiny fraction of the dissolved gold that had been thought, and certainly not enough to cover the massive cost of extracting it.

More recent estimations of the quantity of gold in seawater are more optimistic, putting levels at three times those that Haber considered worth exploring – twenty milligrammes per tonne. In principle, the oceans of the world could contain gold worth some £300 million million at current prices, or to put it another way 400 million Kate Mosses. But even at this more attractive rate, according to Richard Herrington, 'the cost of extraction is too great to be contemplated at the moment'.

There really is, he further notes, gold in the Rhine, 'with production in the best years reaching upwards of 15 kg'.

The sheer unexpectedness of dissolved gold has been successfully exploited on at least one notable occasion. By 1933, Nazi oppression of Germany's Jewish scientists was leading many to emigrate or take refuge in foreign laboratories. Two Nobel physics laureates, Max von Laue, who won the prize in 1914 for his discovery of X-ray diffraction, and James Franck, who won it in 1925 for producing experimental confirmation of the quantization of energy, deposited their medals for safe keeping with Niels Bohr at the Institute for Theoretical Physics in Copenhagen. By the time the German army marched into Denmark in April 1940, Bohr had already donated his own Nobel medal for a war relief auction, but he was concerned to hide the Germans' medals as their discovery in his laboratory would further compromise the already discredited scientists. The medals bore the names of their recipients, and, as they were made of gold, it was illegal to take them out of Germany.

Working with Bohr in Copenhagen was the Hungarian chemist George de Hevesy, who in 1923 had discovered the element hafnium, naming it after the Latin for the city, 'Hafnia'. Hevesy first suggested that they bury the medals, but Bohr felt it was too likely they would be discovered. Instead, as Nazi troops flooded into the city, he set about dissolving them in aqua regia – with some difficulty, he complained later, as there was a considerable amount of gold and it was reluctant to react even with this strong acid. The Nazis took over the Institute for Theoretical Physics and carefully searched Bohr's laboratory, but omitted to enquire as to the contents of the bottles of brown liquid on a shelf, which remained there undisturbed for the duration of the war. After the war, Bohr wrote a letter to the Royal Swedish Academy of Sciences accompanying the return of the medal gold explaining what had happened to it. The gold was recovered, and the Nobel Foundation duly minted new medals for the two physicists.

Aqua regia was one of the many useful, and often unacknowledged, contributions to modern chemistry made by the alchemists, whose discovery that it would dissolve gold naturally occasioned great excitement. In Milton's *Paradise Lost*, Satan is given a tour of the wonders of the earth and sees that the 'Rivers run / Potable Gold'. If solid gold was the symbol of perfection, immortality and enlightenment, its availability in a form that could be imbibed – the solution was typically blended with aromatic oils to make a sort of metallic vinaigrette – surely held promise for a general cure-all.

But gold's other great claim – its resistance to change – left room for sceptics to wonder whether it did anybody any good, or in fact did anything at all. Thomas Browne, the Norwich physician and author, tackles this question in his *Pseudodoxia Epidemica*, an erudite and entertaining catalogue of seventeenth-century urban myths scientifically debunked. 'That gold inwardly taken,' wrote Browne, 'is a cordial of great efficacy, in sundry medical uses, although a practice much used, is also much questioned, and by no man determined beyond dispute.' Observing its 'invincible' passage through fire, he finds it easy to believe gold could also pass through the body without alteration or effect – a thought that prompts him to take a moment to discredit the tales of Midas and the golden goose. But then he changes tack to admit that, though it may not be materially changed, gold might yet exert some effect, perhaps similar to the magnetic force of the lodestone or the electrical charge of amber. In the end he equivocates: 'it may be unjust to deny the possible efficacy of gold'. However, Etienne-François Geoffroy, a French physician and chemist of the following century, had no such doubts. 'Gold,' he wrote drily, 'of all the Metals is the most useless in Physick, except when considered as an Antidote to Poverty.'

I got my chance to try 'gold inwardly taken' one Christmas when I bought some 'gold, frankincense and myrrh' chocolate. The frankincense and myrrh could not compete for flavour with the cocoa solids, but the gold was at least visible as little flakes on each square. I observed no ill effects as I ate. Perhaps it was doing me a power of good, but I felt no elixir boost either. I turned over the wrapper and idly read the list of ingredients. Gold, I was surprised to learn, merits its own E number, E175. It seems that the food regulators, like Browne, want to keep their options open.

## Going Platinum

Wallis Simpson, the twice-divorced American socialite who in 1937 married the former King Edward VIII to become the Duchess of Windsor, was not renowned for her attention to correct social procedure. But on the matter of jewellery she was adamant: 'Any fool would know that with tweeds and other daytime clothes one wears gold; with evening clothes one wears platinum.'

Platinum rose during the first part of the twentieth century to be seen as the preferred jewellery metal of those who found silver simply too common. It is one of the heaviest lustrous metals, fully twice as dense as silver, but not as purely white. It rarely dazzles, but shines with what John Steinbeck called a 'pearly lucence'. Platinum's appeal is particularly relative: it is heavier than silver, trendier than gold. It is fashion's answer to these timeless elements, full of its own importance, and self-ordained in its rank.

At a time of widespread economic suffering, platinum fulfilled the need of an increasingly disconnected high society for a substance more precious, and possibly less obvious, than gold. It is odd in a way, then, that the material chosen for this task is if anything a little more abundant than gold — although both metals are equally scarce in the earth's crust, there is ten times more platinum than gold in the soil. No matter. In due course, platinum — if not the bullion itself, then at least the idea of it as the most valuable of all metals — would percolate down to be understood even among the lower social strata and secure its symbolic place above gold in the league table of luxury. Platinum immediately signified a new kind of rich, a badge of wealth not amassed over ages like a hoard of gold, but acquired suddenly, boldly, speculatively — and liable to be lost the same way. In the second book of his *America* trilogy, *The Big Money* of 1936, John Dos Passos depicts an array of characters as they struggle to reconcile their ideals with the need to get on in the febrile years leading up to the Depression. The 'ghosts of platinum girls' haunt the novel like sirens warning against the temptation of the new riches.

Frank Capra's film of 1931, *Platinum Blonde*, made capital out of the emerging symbolism of the metal, and donated its title to the language in return. The platinum blonde in question is another rich socialite who seduces, marries and then controls a reporter who is investigating a scandal in her family. Jean Harlow took the lead. Originally, the film was to have been called *Gallagher*, the name of the girl who loses and then regains the reporter's affections. But the producer, Howard Hughes, had Harlow under personal contract and insisted on the change of title in order to promote his starlet. It worked, launching both Harlow and a craze for the etiolated hair colour. Through his studios, Hughes even offered a prize to the main-street hairdresser who could best replicate the shade. But perhaps his money was safe: only those who had been on set would have been in any position to know if they had got it right since the film was made in black-and-white.

Platinum was recognized as an element by European chemists in the eighteenth century, hailed then as 'the eighth metal', an exciting addition to the seven known since antiquity: gold and silver, copper, tin, lead, mercury and iron. But it was effectively discovered by the indigenous peoples of South America 2,000 years ago. The native form of the element known as platina – the diminutive of the Spanish for silver, *plata* – occurs as granules or nuggets of largely pure metal with inclusions of other precious metals or iron. It is typically revealed in rivers or during panning for gold when heavy pale grains are seen amid the potentially precious residue after lighter minerals have been washed away. Platinum melts at a temperature far higher than gold, bronze and even iron, and higher than can be reached by charcoal fires. It should have been impossible for indigenous smiths to convert these granules into a form which could then be worked into jewellery and other items. Yet archaeological finds in Ecuador have revealed just such pre-Columbian artefacts, forcing European metallurgists to acknowledge the mastery of the native smiths, who had perfected a method of sintering, whereby a granular material coalesces into a mass without melting, by adding gold dust to trigger the fusion of the metal.

Hell-bent for gold, the Spanish conquistadores had paid no heed at first to the dull grey platina. Some gold mines were even abandoned because the presence of platina rendered them uneconomic. That attitude changed, however, when the work of a young French chemist, Pierre-François Chabaneau, sequestered at the Royal Seminary at Vergara in the Basque country, came to the attention of King Charles III of Spain in 1786. The seminary was in fact something of a mineralogical hothouse and must have concealed quite a hoard of exotic specimens by the time that Chabaneau arrived: the brothers Fausto and Juan José Elhuyar, who had been engaged to teach there, had already isolated the element tungsten from wolframite, an exceptionally dense ore which they had obtained during their studies in Germany. They put Chabaneau to work extracting platinum metal from the raw platina they had accumulated from South America.

In due course, the Elhuyars were promoted to direct the new

mines in the Spanish colonies, while Chabaneau was brought to Madrid and given a luxurious private laboratory in which to carry on his researches into platinum. The king's minister, the marquis of Aranda, saw to it that the state's entire stock of the metal – seen as less valuable even than silver – was turned over to the Frenchman. One reason for the low estimation of platinum at this time was that the Spanish were unable to emulate the New World craftsmen and convert the metal into a malleable form that could be worked into objects. Chabaneau soon thought he had managed to isolate the pure metal, removing the gold, iron and other impurities that made it unworkable. But he was puzzled to find that its properties refused to settle down to a standard pattern (this was because it still contained other, as yet unknown, elements more closely related to platinum such as iridium and osmium). Chabaneau abandoned the work in frustration, but his patron persuaded him to persist. 'Three months later, the Marqués found on a table in his home a ten-centimeter cube of metal. Attempting to pick it up, he said to Chabaneau, "You are joking. You have fastened it down." The little ingot weighed 23 kilograms; it was malleable platinum!'

At first, samples of platinum were passed around among the aristocracy of Europe, with nobody very sure what to do with it. The difficulty of handling the metal meant that it remained essentially useless. (The Spanish crown had learnt the hard lesson that even well-funded scientific research doesn't always yield a quick return on its investment.) The eighteenth-century memoirist Giacomo Casanova records a visit to a lady alchemist, the marquise d'Urfé, who planned to convert hers into gold. Gradually, however, Chabaneau's method caused the new metal to begin creeping up in value. A platinum chalice presented to the Pope by the king of Spain was the first precious object made from his malleable form of the metal. Chabaneau saw that he was in a powerful position and went into business selling platinum ingots, crucibles and other specialist utensils. At the same time, the Spanish government increased shipments of platina from its South American colony of New Granada. In August 1789, a single vessel landed 3,000 pounds of platina. Although the metal was strictly placed under a crown monopoly, it was still cheap enough to appeal to smugglers and counterfeiters who could plate it and pass it off as solid gold because of its comparable density. Spain's brief 'age of platinum' came to an abrupt end with Napoleon's invasion of the country in 1808 and the rise of the revolutionary independence movement under Simón Bolívar in New Granada. Platinum's odd combination of great density and resistance to corrosion made it the perfect choice for casting the standard kilogramme and metre of the French republic, but grander ideas of using it for decorative objects requiring the services of talented craftsmen were soon forgotten.

In the nineteenth century, the price of platinum declined again as new sources were found in Russia and Canada and more economic means were developed for refining it. Russian aristocrats did not find the metal bright enough for their tastes and, in the absence of other demand, Russia in 1828 began minting three-rouble platinum coins in order to make use of its resource. But even this had to stop when the worldwide price of the metal tumbled still further.

Having reached this low point so soon after it was introduced to Europe, how did platinum then rise to overtake gold in value? The law of the markets suggests that if the answer is not to be found in shortage of supply then it must lie with excess of demand. The expansion of technical applications – in electrical equipment and in many industrial chemical processes where the metal serves as a catalyst – is undoubtedly a factor. But more interesting is the *perceived* increase in platinum's value that arose solely for reasons of social status rather than market economics.

In 1898, Louis Cartier succeeded to his father's Parisian jewellery business and made the family name by popularizing the wrist-worn timepiece in place of the pocket watch. Cartier had experimented with platinum for some years and now made the decision to use it wherever he could in place of silver and even gold. The 'white jewels' such as diamonds that were favoured for evening wear ideally required colourless settings. Gold clashed and was seen as vulgar, and silver had a propensity to tarnish. Furthermore, both metals were inconveniently soft. Hard platinum ensured that Cartier's settings, especially of the largest stones, could be made almost invisible and yet still prove highly durable. The slightly grey lustre of the metal compared with gold or silver ensured that attention would be focused on the gemstones alone. Cartier's innovation unleashed a fashion for platinum in the grandest jewellery that lasted until the outbreak of the Second World War when the metal was promptly rationed because of its usefulness as a catalyst in important chemical processes such as the manufacture of explosives. By then, though, platinum had secured a new cachet, capped by the setting of the famous Koh-i-noor diamond in a crown made entirely of platinum for Queen Elizabeth, the wife of George VI, for the coronation in 1937. (Wallis Simpson must have sickened to know her sister-in-law had this bauble!)

As Cartier was changing the rules of high-society jewellery, the revival of the Olympic Games meanwhile implanted the idea of indicating degrees of excellence according to a scale of different metals. The Olympics of ancient Greece had merely awarded laurels to the best athletes. At the first modern Games, held in Athens in 1896, the winner of each event was awarded a silver medal, with bronze going to the runner-up. Only at the St Louis Games in 1904 did the International Olympic Committee decide there should be gold, silver and bronze medals for the first three places, retrospectively amending the medal table for the two preceding Games in line with the new system.

So it has remained ever since. The hierarchy of gold, silver and bronze has become the conventional way to rank performance in sport and the arts. Record companies introduced the gold disc as a way of congratulating their artists – and themselves – when they sold a million copies of a song. Perry Como

was the first international artist to strike gold. When record sales grew and gold discs became too common, the music industry, rather than do the obvious thing and simply raise the sales threshold for gold, saw the marketing advantage of introducing instead the higher tier of a platinum disc in 1976. Under today's rules, an album goes gold when it sells 500,000 units, and platinum at a million. American Express soon followed, trumping its 'gold' charge card with 'platinum' in 1984.

None of this was any longer about the appearance or properties of platinum metal. Nor was it really about its rarity, which is, as we have seen, no greater than that of gold. For most of us – no Wallis Simpsons – platinum's status is the product of a more complicated snobbery. If we perceive it as more desirable than gold, it is entirely by reverse association – by knowing that a record goes platinum after it has gone gold or that a platinum credit card is harder to get hold of than a gold one. In an era when instant coffee, cheap chocolates and lavatory paper are branded 'gold', something had to be found with greater prestige. For now at least, that something is 'platinum'.

## Noble Metals, Ignobly Announced

In April 1803, a small quantity of shiny metal went on sale in a Soho curiosity shop. A leaflet distributed anonymously to London scientists trumpeted it as 'PALLADIUM; OR, NEW SILVER', and promised that it was 'A NEW NOBLE METAL'. It went on to describe the material's properties in some detail: the 'greatest heat of a blacksmith's fire would hardly melt it', for example, and yet 'if you touch it while hot with a small bit of Sulphur it runs as easily as Zinc'.

The announcement caused an instant furore. Who had

placed it? And could it even be true? If it was true, then why had the announcement not been made in the civil spirit of open cooperation that had become the norm in science by this time?

Suspecting a fraud, a talented Irish analytical chemist, Richard Chevenix, visited the shop and bought up all of the substance that had not been sold (three-quarters of an ounce) and immersed himself in a series of analyses to expose the deception. He must have been surprised to find that what he had bought did in fact possess the novel properties claimed for it. Chevenix nevertheless communicated to the Royal Society his opinion that it was not a new metal 'as shamefully announced', and was more likely just an amalgam of platinum and mercury. Other scientists could not confirm Chevenix's result, but scarcely wanted to contemplate the only alternative interpretation – that a major scientific announcement might be made in the guise of an unsigned commercial handbill.

In the end, it was almost that bad. For it soon turned out that the metal really was new to science. Only the fact that the author of the flyer, and of the discovery itself, was one of their own mitigated the disaster: he was the already noteworthy chemist William Hyde Wollaston, who was known to be deeply engaged in a project involving platinum. But why had he conducted himself in such a peculiar fashion in this case?

For fifty years, European governments had eyed the platinum brought from South America with a mixture of lust and despair, aware that it had the potential to be transformed into a lustrous precious metal, dreaming perhaps that it would boost their economies as New World gold and silver had done a couple of centuries before, but lacking the means to effect this transformation. In Spain, Chabaneau had kept his method a close secret and had only found a market for occasional decorative objects.

Wollaston and another chemist, Smithson Tennant, separately took up the problem, and, when they became of aware of one another's interest, decided to go into partnership to see if they could produce Pierre-François Chabaneau's malleable platinum on a larger scale and find new applications for it in science and industry.

Wollaston and Tennant were both the sons of clergymen, and both had studied medicine at Cambridge but then turned to natural philosophy. There, however, the similarities ended. Tennant lost both his parents in childhood and was largely self-taught. Wollaston grew up in a family with fourteen siblings and enjoyed a comfortable path to academic success. Tennant, five years older, was a humorous and kindly man, untidy in his work, often indecisive about his projects, yet always properly observant of the rules of experimental method and reporting when he did finally settle on a course of action. Wollaston was precise and self-controlled almost to the point of obsession – it was said he could write on glass with a diamond a script so small that it could only be read using a microscope. He was also secretive and eccentric and not always easy to get along with. The fruit of their collaboration was to be a substantial fortune from the platinum enterprise and a permanent place in the annals of science, as each man would add two new chemical elements to the thirty-five or so then known. But the way each chose to announce his respective elemental discoveries to the world would reflect their difference in temperament.

On Christmas Eve 1800, the two men bought up nearly 6,000 ounces of river-dredged platina from a disreputable vendor who had probably obtained it smuggled via the British West Indies from New Granada. The purchase cost them £795, a handsome sum, but the quantity was vast, and platinum was as yet far cheaper than gold. If they succeeded in converting this heap of grey crumbs into lustrous metal, they would be very rich men.

Wollaston took the lead in this commercial project, dissolving the raw material a pound at a time in aqua regia, then reacting it with ammonium salts to form a precipitate which could be heated to release the precious metal. However, his ingots turned out brittle and were no use for further work. Tennant meanwhile examined the small amount of black residue that always remained behind when the native platinum was dissolved in aqua regia, quickly becoming convinced that it was not merely graphite as others had supposed, but metallic in nature. By extracting the black powder and carefully treating it with various powerful reagents, he was able to obtain new precipitates of different colours and a pungent oily liquid. These proved to be compounds of two new metals, which Tennant named iridium (after the Greek for rainbow because of the colours of its salts) and osmium (from the Greek for smell). Tennant was closely pursued in this work by French scientists, but had wisely taken the precaution of sharing his hunch that the residue was metallic with Sir Joseph Banks, the President of the Royal Society, thereby ensuring that he was rightfully acknowledged as the discoverer of both elements.

Wollaston followed similar procedures in experiments with the platinum-rich liquor produced by the aqua regia. He too noticed an unexpected precipitate, which he soon satisfied himself contained yet another new metal. He thought to call it ceresium after the minor planet Ceres, which had been discovered a few months before, but then opted for the name palladium. However, rather than publish or communicate the news of his discovery informally as Tennant had done, Wollaston waited until he had amassed a significant quantity of the new metal – and then made his eccentric decision to advertise it for sale in small portions charged at five shillings, half a guinea and one guinea.

PALLADIUM;

OR,

*NEW SILVER,*

HAS these Properties amongst others that shew it to be

A NEW NOBLE METAL.

1. IT dissolves in pure Spirit of Nitre, and makes a dark red solution.

2. Green Vitriol throws it down in the state of a regulus from this solution, as it always does Gold from *Aqua Regia.*

3. IF you evaporate the solution you get a red calx that dissolves in Spirit of Salt or other acids.

4. IT is thrown down by quicksilver and by all the metals but Gold, Platina, and Silver.

5. ITs Specific Gravity by hammering was only 11.3, but by flatting as much as 11.8.

6. IN a common fire the face of it tarnishes a little and turns blue, but comes bright again, like other noble metals on being stronger heated.

7. THE greatest heat of a blacksmith's fire would hardly melt it;

8. BUT if you touch it while hot with a small bit of Sulphur it runs as easily as Zinc.

IT IS SOLD ONLY BY

MR. FORSTER, at No. 26, GERRARD STREET, SOHO, *LONDON.*

In Samples of Five Shillings, Half a Guinea, & One Guinea each.

J. Moore, Printer, Drury Lane.

When Chevenix communicated the results of his investigation, it put Wollaston in a quandary. He was now unable to claim the discovery that was his due without admitting his subterfuge. Instead, he issued another anonymous communiqué, this time in a chemical journal, offering a reward of £20 to the person who, before a jury of three chemists, could make twenty grains of palladium. Nobody seems to have risen to the challenge. Meanwhile, he quietly went on with his researches. His next discovery was to offer him a way out. Further experiments with the raw platina and aqua regia yielded new rose-coloured salts, indicative of another new element, which Wollaston named rhodium. This time, there would be no foolishness about the announcement. His friend Tennant had recently given his own paper, formally announcing his discovery of iridium and osmium. Wollaston followed his example and read out his paper on rhodium to the Royal Society in June 1804. He did not use the occasion to disclose the mystery of palladium, but a few months later, he wrote again to the journal where he had advertised his reward, explaining that it was he who had secretly discovered palladium and offered it for sale, putting forward as excuse for his behaviour that chemical anomalies observed at the time of the palladium discovery had prevented him from announcing it then, and that he had resolved these anomalies with the subsequent discovery of rhodium. This was not quite true, but it did enable Wollaston to save face.

The new elements at last explained the brittleness of the platinum ingots. Armed with this knowledge, Wollaston pressed on with his manufacturing process and eventually obtained a valuable product. Over the next fifteen years, he built a tidy business constructing platinum boiling vessels for use in chemical factories and other specialist devices. Wollaston only revealed the details of the process a month before his death in 1828 when he knew he was suffering from a fatal illness.

Over the years, Wollaston and his partner Tennant purchased some 47,000 ounces of native platina and produced 38,000 ounces of malleable platinum – about a bath full – as well as 300 ounces of palladium and 250 ounces of rhodium, enough to fill a pint glass with each metal. Some of the platinum was formed into crucibles for scientific experiments or rods for drawing into wire, but most of it went to gunsmiths, who used the metal to improve the contact points of flintlock pistols where it was cheaper and more effective than the gold they had been accustomed to using for the purpose. Wollaston and Tennant bought their platina at a typical price of two shillings for a thousand ounces, and sold pure platinum at sixteen shillings an ounce – a 6,000-fold increase! The secrecy that necessarily attended the perfection of this highly lucrative process simply appears to have warped Wollaston's judgement when it came to the discovery of palladium.

Wollaston's career prospered nevertheless. His momentary lapse of scientific protocol was forgiven, and he won admiration for further discoveries in chemistry and optics as well as for his platinum process, from which he may have made a fortune of £30,000 or more, equivalent to quite a few million pounds today. Chevenix, disheartened by the episode, renounced science, married a French countess and turned to writing historical dramas.

## The Ochreous Stain

Earthly power may arise from the possession of gold, but iron once radiated celestial power. Lumps of it fell from the sky – they still do. These iron meteorites, gifts of pure metal handed down from the heavens, held an instant sacred appeal. In some ancient beliefs, the sky itself was made of metal. Ilmarinen, the Eternal Hammerer of Finnish mythology, was said to have hammered out the firmament at the dawn of time. A myth for a grey-skied land.

Having fallen from the sky evidently directed by nothing other than divine will, these aeroliths represented heaven on earth more satisfactorily than any terrestrial material or artefact sanctified by man. Worship must have begun long before it was possible to think of working the metal: there would have been little else to do with the mysterious burnished masses than to place them in the temples. But in more technological times, iron also throws down a moral gauntlet. According to the Qur'an (Sura 57:25), God sent down messengers, scripture and law; 'And We sent down iron in which is mighty war, as well as many benefits for mankind, that Allah may test who it is that will help, unseen, Him and His messengers.'

The Hayden Planetarium at the American Museum of Natural History in New York is home to some of the largest iron meteorites ever found. One prize is the Willamette meteorite, a fifteen-tonne black-and-silver lump the size of a small car and contoured like a piece of popcorn. It is almost pure metal – iron with a few per cent of nickel – polished by the touch of visitors over the course of the century that it has been on display. Visiting the museum one day, I find it crowded round by children like a tree in a playground. I touch the meteorite, and realize that I have done so entirely casually. I feel no magic – unlike the time when, at another museum, I was lucky enough to be allowed to hold in my hand a tiny meteorite that had fallen to earth having first been blasted off the surface of Mars. Other visitors touch the Willamette iron too, with curiosity and admiration, with rude familiarity or casual indifference, but not with special reverence. Paradoxically, it is the museum setting that makes this remarkable object seem ordinary, just one among hundreds of spectacular exhibits. I struggle to reimagine the metal mass lying in the crater of its own making deep in the Oregon forest where it was found. There, it could only look alien, truly an object from another world, a gift from the gods.

The meteorite was found by chance in 1902 by a Welsh immigrant, Ellis Hughes, on land belonging, fittingly perhaps, to the Oregon Iron and Steel Company. Over a period of months, Hughes dug out the massive lump, constructed a cart and transported it the short distance to his house. Claiming to have found the meteorite on his land, he charged people twenty-five cents a time to see the curiosity. Unluckily, one of his visitors was the lawyer for Oregon Iron and Steel, who suspected that the iron had been taken from company land. Hughes duly lost the complicated legal case that followed, and the company gained possession of the meteorite, later selling it to the donor who gave it to the museum.

The Willamette meteorite is deeply pitted from centuries of corrosion in the humid woods. The best iron meteorites tend to be found near the poles, where they are preserved in ice. In 1818, the British Arctic explorer John Ross was surprised to come across Inuit hunters using steel tools. He suspected their metal was meteoritic in origin, but it was not until 1894 that an American expedition led by Robert Peary found the source – three of a group of meteorites that had been named by the Inuit according to size: 'tent', 'man', 'woman' and 'dog'. With a great effort, Peary retrieved the thirty-one-tonne 'tent', which is also now in the American Museum of Natural History along with 'woman' and 'dog'. The fourth of the group, 'man', was only found in the 1960s and was taken for display in Copenhagen.

There is a delicious irony here: in order to recover the massive iron meteorites that he found in the Arctic ice Peary was obliged to build a railway. Its construction must have required the import of a quantity of iron far in excess of the mass of the meteorites – proof that celestial iron retains its power over the terrestrial.

Iron meteorites made mighty objects of worship. But where sheer survival was the priority, the practical value of the metal could not be ignored. For a long time before it was discovered that it could be extracted from terrestrial ores, this metal from heaven was humankind's main source of iron. Meteorites fall rarely, however, and so in societies from ancient Egypt to the Aztecs, iron was appreciated for its utility, yet at the same time often regarded as more precious than gold. Objects forged from it such as swords were functionally superior to any alternative. Some Bedouin believe that a man armed with a sword of meteoritic iron becomes invulnerable and all-conquering – something quite plausible given the superior qualities of the alloy. But the raw material was never abundant enough to equip armies, and so these weapons were reserved for ritual rather than practical use. The folk memory of a time when forging iron meant working with material from heaven begins to explain the mythic potency of iron and the smiths who have mastery over it.

It was around 5,000 years ago, probably in Mesopotamia, that humankind gained the ability to smelt iron from widespread terrestrial ores. Gradually, reverence for these celestial objects was replaced by sheer incredulity. Well into the nineteenth century, even the most learned societies scorned the idea that lumps of pure metal could simply drop from the skies. On one occasion, the French Academy of Sciences passed a vote that iron meteorites did not exist. Only later were new techniques of analysis able to confirm their other-worldly nature. Specifically, iron meteorites tend to contain a significant proportion of nickel, which indicates that they cannot have been made from terrestrial ores – they are in effect a kind of stainless steel. Indeed, when alloy steel was first produced with nickel it was marketed in recognition of its superior properties as 'meteor steel'. Conversely, if nickel is absent from the iron in an ancient object, this tells the archaeologist that the iron must have been smelted from ore.

Although the *words* for all the metals in languages derived from Latin are gendered male (neuter in German), it is quite clear that the substances themselves connote gender quite independently of this linguistic happenstance. Gold and silver are linked with the sun and moon, which are almost universally regarded as male and female. In Greek mythology, for example, the sun god Apollo is clothed in gold, and his sister Artemis hunts with a silver bow, and for the Incas the moon was the incestuous bride of the sun. Other ancient metals may be more ambiguously gendered – mercury, for example, is the female principle to sulphur's male in Chinese and Western alchemical theory, yet is linked to the male god Shiva in Hindu tradition. However, no metal is more clearly masculine than iron.

When the Soviet press called Margaret Thatcher the Iron Lady for her persistent opposition to communism she took it as a compliment. Iron has always indicated strength and toughness – qualities almost synonymous in everyday usage, but which have rather precise meanings in materials science. The metal is generally hard, which means that it changes shape only very slightly when large forces are applied to it, but it is also less ductile and malleable than the other ancient metals. It is this unbending quality, not simply its hardness, that drives 'iron' as a metaphor. Churchill's inspired coinage of the 'iron curtain' draws on this physical and attitudinal inflexibility, as well as making sly reference to Stalin, a *nom de guerre* meaning 'steel'. Wellington, on the other hand, earned his nickname as the Iron Duke not through military prowess but for putting iron shutters on the windows of his London home as protection against 'the mob'.

Iron's masculinity is reinforced by the metal's eminent suitability for making weapons of war. However, this is not to say that fashioning a serviceable sword was an easy business. At Sutton Hoo in Suffolk, an Anglo-Saxon royal burial site uncovered in 1939, archaeologists found the helmet, made from a single piece of iron, thought to have belonged to King Raedwald, who died around 625 CE. They also found his sword and shield, though less well preserved. The blade of the sword was pattern-welded, a process that involves layering sheets of iron together to build up the shape of the blade, and which often results in a fine decorative pattern at the surface. In this way, desirable properties can be directed where they are needed – extreme hardness towards the edge of the blade, but a degree of flexibility in the core so that the weapon does not shatter upon impact. The skill of the forger lay in his intuitive knowledge of when to incorporate more carbon, obtained from the charcoal of his fire, into the molten iron to produce a harder steel. The Sutton Hoo visitor centre has arranged a display of iron sheets and rods of the kind that the swordsmith would have started with. They look like new grey plasticine. But without the heat of the forge, I find it hard to comprehend how they might be transformed into such a beautiful weapon, or to sense the patient, repetitive actions of heating and softening, hammering and quenching, with their implied cycle of death and rebirth by fire, that would have endowed the sword with ritual significance.

The long-time rarity of iron and the technical difficulty of forging it made blacksmithing a trade of great prestige and mystique. The smithy was a place of hellish fire and stench from the sulphur released by unconverted ore. Wayland or Wieland, the blacksmith god of the Anglo-Saxons, like Hephaestus in Greek myth, is often represented as banished along with his forge to an island because his work is so repulsive. Yet the smith himself is the master of a necessary art and is known for his ingenuity as well as his skill. Ilmarinen, for example, in Finnish mythology, is an inventor as well as a smith.

Swords made of iron were thus exceptionally precious artefacts – far too precious for use in real battle – and it is natural that they were seen as possessing mythic qualities. Although the metallurgy of these weapons is not always explicit, it seems that Excalibur, the sword of Arthurian legend, was made of iron – the name may derive from the Welsh *caled*, meaning hard, or from the Greek and Latin word for steel, *chalybs*. Sigurd's sword, Gram, in Norse mythology is iron too. Iron craftsmanship has been raised to a high art in Japan, whose islands are poorly supplied with copper for bronze, or with the metals regarded as precious elsewhere. Kusanagi, the seventh-century sword that is part of the imperial regalia of Japan, is thus almost certainly made of iron, although it is impossible to know as the object, or its replica, is kept in a shrine where it is forbidden to be inspected.

Not content with the scene in *The Ring Cycle* where the hero Siegfried forges such a magic sword, Wagner also began an opera based on the legend of Wieland the Blacksmith (as well as another based on E. T. A. Hoffmann's story 'The Mines of Falun', set in Sweden's vast copper fields, which we will hear more of later). When excerpts from documents supposed to be Hitler's diaries were published and then sensationally revealed to be forgeries in 1983, it was one of the more plausible aspects of the hoax that Hitler, an avowed Wagnerite, should have taken up the unfinished task.

Though iron has long been accorded warlike male attributes, it is only with the advent of modern scientific methods that it has been possible to prove that the red of blood and of iron ore are due to the same cause. Yet the connection seems to have been sensed long before. When Siegfried slaughters the dragon Fafner with the sword he has made, he licks the dragon's blood that has spilt on to his hand. The blood, like the sword, confers magical powers, and the hero is suddenly able to understand the birds in the forest. Perhaps even Irn-Bru – 'made in Scotland from girders', according to the advertising – appeals in part because it flirts with the taboo against drinking blood, although the amount of iron it contains is minuscule, and its rusty colour is mainly due to E numbers.

Though noted often enough, the metallic taste of blood was only explained in the mid eighteenth century. It is a story seldom found in histories of science. Yet the experiment was simple, and seems to have been first performed by Vincenzo Menghini, a Bologna physician, around 1745. He roasted the blood of various mammals, birds and fish as well as humans. He then poked among the solid residue with a magnetic knife and was pleased to find that particles of it were picked up on the blade. From five ounces of dog's blood, he obtained nearly an ounce of solid material, the bulk of it being magnetic. (He presumably obtained similar results using human blood, although accounts do not explain how he got hold of it.)

The experiment is very easy to repeat: place in a ramekin a tablespoon of blood (I drained mine off a pack of frozen chicken livers) and partially evaporate in a low oven. Transfer the sludgy residue to a small crucible or other container able to resist heat and roast to dryness. Scrape out the residue and grind to a coarse powder until it resembles coffee grounds. Spread the powder on a sheet of paper, and pass a moderately strong magnet closely over it. A few particles will be lifted on to the magnet.

This was clearly the result Menghini had been expecting. But the question then arises: why did he think iron would be present? It can only be because the association of iron and Mars, blood and war, originating in Greek and Roman mythology, was so securely rooted in the alchemical orthodoxy of the time, even to the extent that those suffering from disorders of the blood were sometimes recommended to take iron salts. Further evidence that it had long been tacit knowledge that iron and blood were connected in some way comes from the name of one of the metal's principal ores, haematite, a sixteenth-century coinage, the prefix haem- being derived from the Greek for blood.

Menghini too went on to make up preparations rich in iron which he fed to human and animal subjects, afterwards observing the enrichment of the red blood cells, thereby proving that the colour was linked to iron. His research made a vital contribution towards explaining – and curing – chlorosis, a disease characterized by a greenish pallor of the skin, which only then acquired its present name of anaemia, from an- plus haem-, meaning without blood.

Iron's association with Mars has equally confused beginnings.

It was natural enough for mystics and philosophers to seek correspondences between the sun and the moon and the five observable planets and the similar number of ancient metals. But in the absence of a competent metallurgy it was impossible to decide which metals were pure and irreducible rather than mixtures. As a consequence, brass, bronze and alloys used for coins were often placed on an equal footing with gold, silver, lead and tin, while the special alchemical status of mercury meant that it was not at first linked with a planet at all. In Persia, iron was first linked with the planet Mercury. Only much later did Western alchemists reassign Mercury to its element namesake, freeing iron to pair with Mars.

When was it first thought that Mars might have a more material connection with iron? The invention of the spectroscope in 1859 enabled scientists to analyse the light emitted by luminous bodies, leading to the discovery of several new elements identified by the signature colours of their flames. A spectrum is like a rainbow in which only a few bands of colour appear. Each element has a characteristic atomic spectrum, due to the absorption and emission of light associated with the unique energy levels of its orbiting electrons. These early spectroscopes, however, were only sensitive to emissions of light, such as from laboratory flames or from the sun. They were not able to say anything about the light reflected from non-luminous objects that gives them their colour. Scientists might speculate that the red planet was rich in iron ore, but it was no more possible to check this than it was to prove that the moon was not made of cheese. And at the time when they might profitably have begun to investigate the question, in the last years of the nineteenth century, many were in any case distracted by the planet's white, earth-like poles and the supposed 'canals' crisscrossing its surface.

It was not until spacecraft – *Viking* in 1976 and *Pathfinder* in

1997 – landed on Mars that the origin of the colour could at last be explained. Rather than the dark blue expected from its thin atmosphere, they found the sky to be the colour of butterscotch owing to dust storms. The planet's surface is covered with the same fine dust, composed of the iron oxide mineral limonite. Recent analysis of data from the Mars landers has indicated that the concentration of iron on the planet's surface is greater than it is in the crust below, suggesting that the iron may have originated from meteorites rather than as a result of volcanic eruptions bringing mantle rock to the surface.

It is rare that science finds itself in a position to vindicate superstitious belief, but it happened twice with the revelation of iron in blood and on Mars.

Iron today brings to mind not venerated meteorites or magic swords, but the engineering achievements of the Industrial Revolution. The Romans had made good use of the metal for weapons, tools and construction, but it was not until 1747, when it was found how to use coal with iron to make steel, that the metal really took over. In that year, Richard Ford, who had inherited Abraham Darby's pioneering iron foundry at Coalbrookdale in Shropshire, showed that it was possible to vary the amount of coke or coal added to the ore in order to produce iron that was either brittle or tough. The greater control of the metal's properties achievable by the addition of small amounts of this carbon allowed iron to be manufactured for very different uses, from the structural beams of great bridges to the cogs and wheels of steam engines and spinning machines.

The most transforming, extravagant and joyous expression of the new iron age was the railroad, an innovation whose debt to the metal is recorded in practically every language except English: *chemin de fer*, *Eisenbahn*, *ferrovia*, *vía férrea*, *järnväg*, *tetsudou*. The iron way swiftly made this element a more visible symbol of power than gold ever had been or silicon ever would be. Sentimental poets naturally read the Industrial Revolution as a destructive force and used its iron as a chief sign of its enslaving effect. As early as 1728, James Thomson, the Scot responsible for the words of 'Rule Britannia', bewailed the loss of the poetic golden age in 'these iron times'. Blake's long poem 'Jerusalem' positively clanks with such references, as in this sharp tirade against both science and the technology to which it gives rise:

> O Divine Spirit, sustain me on thy wings,
> That I may awake Albion from his long and cold repose;
> For Bacon and Newton, sheath'd in dismal steel, their terrors hang
> Like iron scourges over Albion.

But it wasn't all bad. I think Aldous Huxley struck nearer the mark, when, in *Eyeless in Gaza*, he comments as his central character embarks with childlike delight upon a train journey: 'The male soul, in immaturity, is *naturaliter ferrovialis*.' (That is to say: boys by nature love railways. Huxley's typically clever allusion is to the early Christian writer Tertullian's belief that the soul is by nature Christian, *anima naturaliter christiana*.) Roman iron might have been the material of shackles and chains, but Victorian steel opened up new territory, crossed oceans and brought people into contact; it literally built bridges. The magnificent cast iron bridge thrown across the Severn near Coalbrookdale in 1779 is now a UNESCO World Heritage Site. The Menai Strait suspension bridge designed by Thomas Telford in 1819 used wrought iron chains to span a channel 166 metres wide. This met the requirement of the British Admiralty that shipping be free to pass underneath, something that would have been impossible with a bridge based on stone piers. Thirty years later, Robert Stephenson completed a second iron bridge based on a tubular box design that would carry the heavier load of a steam locomotive across the strait in a rectangular tunnel. Between them, the two structures demonstrated the lightweight structural gymnastics that were possible with properly engineered iron. From Joseph Paxton's Crystal Palace to Isambard Kingdom Brunel's *Great Eastern*, we still regard these engineering achievements with real wonder. But the railway above all occasioned excitement at the time – think of Turner's tumultuous painting of a train rattling along a viaduct, *Rain, Steam, and Speed* – and still occasions fondness in the memory.

As the iron meteorites that fell to earth now show, where there is iron, rust is never far behind. Rust has its own potent symbolism, linked to its distinctive bloody colour, and proportionate to the power of iron. As the rise of the industrial age had been accompanied by images of fresh-forged iron so its decline was to be streaked with rust. The band of American states from Michigan east to New Jersey became known as the rustbelt as their steelyards and metal-bashing industries succumbed to foreign competition. The image of rust might be expected to be an entirely negative one. But not so. Just as the love of ruins stems from the vicarious thrill of imagining the collapse of our own civilization, so the corrosion of iron and steel back to the more natural form of rust appears to promise an Arcadian return. Even at the height of the Industrial Revolution, John Ruskin longed to see time and entropy do their work. In 1858, he gave a lecture at Tunbridge Wells, where the famous spring water was apt to turn rusty, commending the 'ochreous stain' which he said should not be seen as 'spoiled iron' but as the element in its 'most perfect and useful state'. (He ignored an obvious tautology for the sake of his happy phrase, since ochre is simply iron oxide anyway.)

Ruskin's sentiment has been enthusiastically endorsed by modern sculptors whose preference is often for steel with an instant patina of rust. Antony Gormley's *Angel of the North* in

Gateshead embraces multitudes with its wide metal wings. The steel from which it was built recalls the heroic shipbuilding for which Tyneside was famous (ironically from around the time of Ruskin's lecture), but the rust plainly records its demise. Richard Serra's great arcs of rusted steel, too, strike me as salutary reminders that our feats of accomplishment are only transient. Most squat in galleries and city squares, but at the Louisiana Museum outside Copenhagen, I discover another Serra slab spanning a wooded gulch. It is a kind of inverse of the achievement of that first great iron bridge – a valley blocked, not crossed, its iron not preserved from nature but left to decompose quietly into the fallen beech leaves. I walk up to the brown wall and tap it to reassure myself that metal lies underneath. I rub my fingers along it as Ruskin must have done to some already neglected Victorian device to capture the ochreous stain. The colour tastes of blood. I wonder if the Martian meteorite I'd held would taste the same – the taste of human blood in a stone from Mars generated from celestial iron.

## The Element Traders

The starting point for this book was my adolescent collection of the elements. I probably never got past thirty or forty out of the hundred-plus complete set, harvesting them from round the house, even with the help of one or two more elusive substances stolen from school. I am not a natural collector. But as I set to work this time, I begin to realize there is quite a community of people out there who have stuck with it and who have not only completed their set, but made it into a project, a mission and even a business.

They are abetted in this by the internet. The periodic table provides the perfect map, a familiar visual mnemonic that leads down many rabbit-holes. Peter van der Krogt, a geographer at the University of Utrecht and a cartographic historian, clearly appreciates this. His website gives the etymology and the history of the discovery of 112 elements. (The site also includes a link to his collection of car licence plates and coins with maps on them.) On another website, Theodore Gray's periodic table is a masterpiece of the carpenter's art – he will even sell you a periodic table table. The story of each element lies on the other side of an engraved wooden portal. Once past this timber threshold, there are beautiful images of the element and its minerals and details of where and how he obtained them. The sources are sometimes exotic, but more often very ordinary: his cerium comes from a camp-fire starter bought from Walmart, his bromine in the form of sodium bromide used to salt the water in hot-tubs. He also accepts donations. 'A lot of people seem to have an element or two in their attic,' he notes laconically on the site. 'By the way, if you have any depleted Uranium from Afghanistan, I could use it.'

Max Whitby and Fiona Barclay have made more than a hobby of the elements. They are element traders, supplying fellow enthusiasts like Gray with specimens of the pure elements from their mews studio-cum-laboratory in a former chocolate factory in West London. Whitby is a former director of the BBC's *Tomorrow's World* programme. He set up a business publishing multimedia before going back to school and rediscovering his scientific roots. He has recently been awarded a doctoral degree for research in carbon 'nanopipes', the tiny rolls of graphite that are currently one of the hottest fields of chemical investigation. Fiona manages a company called Bird-Guides, which produces exactly what its name suggests. Together the two have combined their interests to produce lavish natural history DVDs, and, on the side, run their trade in the elements.

We meet for lunch at a local greasy spoon serving improbably authentic Thai food. Max and Fiona have come prepared. Out on to the table come samples of various metallic elements, solid lumps the size and shape of thirty-five-millimetre film canisters. I am invited to guess what they are. Magnesium and tungsten are easy enough to tell apart, but others have me stumped. So many are similar at first glance, with the same grey sheen. However, closer inspection reveals slight differences. The light they reflect is subtly different in colour – some metals have the barest tinge of pink, or yellow, or blue. The surfaces have all been polished smooth, but according to the way the metals naturally solidify, they vary in appearance, some being almost mirror-like while others are slightly grainy, hinting at a distinctive crystalline microstructure.

It is when you pick up the specimens that they really start to separate themselves. You start out with a fairly clear idea of what you think a lump of metal that size should weigh – information you've gathered over a lifetime from shuffling coins or handling kitchen utensils. But these more unusual specimens soon confound those expectations. Some, like the tungsten, are astonishingly heavy, and a few are so improbably light that you doubt they can be metal at all and wonder if they are cunningly disguised plastic. Lifting them in turn, you unlearn what you think substances ought to weigh and learn to be surprised each time by how heavy or light each sample is compared with the last. They feel different against the skin. Some are warm to hold, others seem to suck the heat from your hand. They smell different too, some metals tainted by the grease of previous touchings, others maintaining a citrus cleanness. As I pick up one sample after another, I am disappointed at how many I get wrong. I take some consolation from the fact that the bar is set pretty high. One specimen is a block of hafnium, an element mainly used to make control rods in nuclear reactors. What on earth are they doing with it? 'Contemplating it,' says Max.

Why the elements, I ask. 'I like the way the table explains our world. Each allocation is a little bit of our civilization,' Max suggests. For Fiona, it's about collectible sets – 'birds, butterflies and elements'.

The trick is to buy the elements in bulk, as they are typically supplied for industry, and then to melt them down and remake them into more attractive forms. Most enthusiasts prefer their metal elements prepared as shiny beads that show their lustre to good effect. Others, German collectors especially, for some reason, want more natural-looking specimens, and creating these may involve heating and cooling pieces of the element in such a way that they form large crystals.

Perhaps thirty of the elements can be bought over the counter, if you know the right counter. Magnesium, for example, is sold by ships' chandlers for use as 'sacrificial anodes' placed below the waterline where they corrode in preference to other metal parts of the vessel. Max's raw magnesium is the sacrificial anode of an oil tanker, a whopping lump the size of a hip bath. Rarer metals used as catalysts are sold in the form of powder. Max and Fiona chop and mould these raw materials into the prettier forms that customers appreciate as 'elements as they truly are'. Whether this is as they truly are is a moot point, of course. But diced, sautéed and served up in these ways, the elements are certainly deliciously transformed. The inert gases come in discharge tubes bent into the shape of the letters of their respective chemical formulae. The most reactive or poisonous elements come in sealed ampoules – subject to shipping restrictions. Even radioactive rarities such as radium and promethium are offered for sale, in the form of glowing hands salvaged from old wristwatches and safely encased in resin.

Their clients include schools and chemical companies, for whom they construct beautiful displays of elements and their compounds pigeonholed into illuminated boxes. But another significant portion of their business comes from obsessive individuals. Radiologists are prominent among their customers: perhaps the dependence for their work on the ability of radioactive forms of certain elements to decay into other elements leaves them craving the apparent fixity of the periodic table. For others it is undoubtedly the finitude that appeals. A complete set of the elements is after all the ultimate collection: from it you could in principle make anything found in any collection anywhere.

They show me beads of rare metals: rhodium, ruthenium, palladium and osmium. All these elements are closely related to platinum and share its serious grey lustre. They look extremely similiar, although detailed examination reveals slight variations among them. I can see that osmium, for example, has a distinctly blue tinge compared with its precious neighbours. I heft the pieces in my hand – the densest of all the elements and therefore the densest substance known to exist. I cautiously sniff them too. Although osmium metal is benign, its volatile oxide is one of the smelliest and most poisonous substances known. I am relieved to find I can smell nothing. In 2004, osmium tetroxide was at the centre of a terrorist alarm in London. I ask if innocent element-trading isn't made difficult because of such things. Max admits he has been visited once or twice by the 'nuclear police'. 'They were very nice. They gave us some advice on how to improve our inventory.'

Max and Fiona's task for the day is to prettify some industrial molybdenum. Molybdenum is a good example of the many elements we tend to hear little about even though they are not rare and are often quietly useful, this one being employed mainly in specialist steel alloys. They start with a few pieces of dull grey metal in powder form pressed into cake rather than cast ingots or forged bars. Molybdenum has one of the highest melting points of all the elements, and so the next stage is quite a palaver, requiring a powerful electric furnace. The floor of the furnace is a copper plate which will itself be prevented from melting under the extreme heat by cold water running underneath it. Around this is what looks like a glass bell jar, but is in fact a protective screen of quartz, which makes a transparent circular wall. The whole contraption is no bigger than a pressure cooker, but seems able, like an Elizabethan theatre, to contain worlds.

Unexpectedly, there are three chemical actors upon the copper stage: small pieces of tungsten and titanium as well as the molybdenum. Fiona opens the valve on a nearby gas cylinder that pumps inert argon gas through the chamber. Max switches on the current – 453 amps, fed from a thrumming electric welder of the type used in building steel bridges. The tungsten – the only electrical conductor that would not melt – serves as the 'striker' that will complete the circuit and ignite the flame. Next, the small piece of titanium is sacrificed in what feels like a ritual but is simply a precautionary way of mopping up any oxygen remaining in the chamber which might otherwise spoil the molybdenum. Then Max brings the flame to each chunk of grey molybdenum in turn. Viewing the proceedings through a thick plate of dark glass, I see the metal glow orange and pucker up into a bead. The orange fades as each bead cools before, miraculously, a bright gleam seems to force its way through the sooty surface. The three elements have all responded differently to their shock – one transformed, one destroyed, one unmoved. The drama is complete. When they have cooled, Max trickles the shining beads of molybdenum into my hand, dimpled like overcooked peas. They are brighter than iron and a little greyer than chromium. I tuck them away to add to my own periodic table.

## Among the Carbonari

As long ago as 1939, a man styling himself 'The Last Charcoal Burner' was said to be making his living by supplying the grill rooms of London hotels. Yet he was not the first pretender to this title, nor the last. Obadiah Wickens of Tonbridge in Kent and Harry Clark of East Sussex each purported to be the last before him. And in the Forest of Dean, Edward Roberts, who had been calling himself the last charcoal burner as early as 1930, was still plying his trade in the 1950s. Perhaps it is long hours spent pondering the stifled flame of their fires that inspires such doomy claims.

These days, I am able to find a charcoal burner without difficulty. It would even be straightforward to track one down in my own thinly wooded county of Norfolk, but instead I have chosen to pay a visit to Jim Bettle, who works in the woods of Blackmoor Vale, where Thomas Hardy set *The Woodlanders*.

This book has indelible, though not exactly fond, memories for me, being the novel I was required to study for O Level. Jim picks me up near his home in Hazelbury Bryan, and we drive for a few miles before turning up a hillside track and passing through locked gates on to private roads until we reach the wood where one of his kilns should be ready for emptying.

In a late addition to the preface of *The Woodlanders*, Hardy responded mischievously to the many enquiries he had received from readers regarding the location of 'Little Hintock', the village where the action of the book takes place. Even he did not know quite where it was; he said: 'To oblige readers I once spent several hours on a bicycle with a friend in a serious attempt to discover the real spot; but the search ended in failure.' Although the academic wisdom has it that Little Hintock is based on a village called Minterne Magna some miles to the west, Jim has reason to believe that the place is in fact Turnworth, which is the hamlet nearest to where we are heading.

Unlike Hardy's woodlanders, who were obliged to scrape a livelihood from the living fuel that grew around them, Jim turned to charcoal-burning by choice. Having seen that the local golf courses and estates habitually just burnt wood off as waste, he felt he could do better, and began to investigate potential markets for locally prepared charcoal. In 1996, he bought his first kiln and went into business. Jim recounts a conversation with his Business Link advisor, who was full of admiration for his ambition, but spoilt the effect somewhat when, after an hour's discussion, she asked him where he was going to be digging up his charcoal. 'It's amazing how many people don't know that charcoal is wood,' he says. Charcoal is almost pure carbon – purer than most coal – and when efficiently burnt releases more heat than wood burnt in an open fire. It is also largely lacking in the sulphur and oils that make coal so unpleasant.

We arrive at our destination, Bonsley Wood, high on the hill south of Blandford. Jim's kiln is a steel drum two or three metres in diameter covered by a thin steel lid. Round its edge are eight little hatches that control the rate of burning once the fire is set. It sits harmoniously enough in a hazel clearing, its rust walls blending with the autumn colours. Jim and his helpers will typically set up a kiln where the surrounding wood to be coppiced – undergrowth of hazel, birch and ash mostly – will support a dozen or more burns. Then they will move it to another location. They do this two or three times during the season with each of their kilns. It is mid October when we meet, and this particular kiln will now be put to bed for the winter; the 135th burn of the year. Other woods are burnt to make charcoal for specialist markets: artists favour willow charcoal; laboratories, which use charcoal as a neutral absorbent, prefer pine. Manufacturers of pyrotechnics buy several different charcoals to give their explosive mixtures just the right oomph.

Each kiln has capacity for a tonne and a half of wood but will yield only a quarter of a tonne of charcoal. This simple fact of charcoal's physical nature immediately explains the itinerant life of the charcoal burner. It is far more efficient for him to burn the wood where the wood grows than to transport it for burning in some remote permanent kiln. This in turn confers his marginal position in society, a man apart from the community, always on the move, hidden by trees, and perhaps of no fixed abode.

The wood is carefully arranged for each firing. First, a core of charcoal from the previous burn is piled up in the centre. Long timbers called runners are then laid out from the top of this heap towards the vents to ensure an airway to the heart of the fire. After that, other timbers are carefully layered in, interleaved with more charcoal. Smaller pieces of wood are placed towards the rim of the kiln, and larger ones at the centre where it is hotter, so that all the wood burns evenly. Although Jim's kilns are steel, this careful selection and arrangement of the wood and charcoal is part of the traditional method of charcoal-burning that goes back to ancient times, when the wood was stacked in a shallow pit dug into the ground and then covered with turf to control the rate of burning.

The fire is lit by igniting the charcoal at the centre and allowed to flare up before the steel lid is put on. This restricts the amount of oxygen let into the kiln and prevents the carbon in the wood from being consumed altogether and converted into carbon dioxide gas. From now on, there is no flame and very little smoke as the wood is carefully burnt to charcoal. The rate of burn is governed by the eight vents around the foot of the drum, which are alternately fitted with long chimneys so that they act as flues, or left alone to serve as air intakes. Jim and his men swap the chimneys round during the course of the burn to ensure that all the wood inside the kiln receives equal heat.

At the kiln we are to empty, the fire was set two days ago and

then starved of all air from the following morning, which has given it twenty-four hours to cool. Jim and a helper lift off the lid. The charcoal is not all black as I had expected. Freshly prepared, it lies in large smooth limbs with a sheen like brushed steel. Many pieces still hold the shape of the branch or trunk of the tree that went in. In some cases, it is almost possible for me to identify the species of timber. The job is simply to reach into the kiln and lift out the pieces. Twisted in the hands, they break into fragments suitable for bagging as barbecue fuel. The charcoal is indeed surprisingly light – I find it takes many gathered handfuls to fill a ten-kilogramme paper sack.

It may be too much to claim that charcoal-burning is undergoing a revival. There are only a few score men like Jim in the country. 'It's quite challenging to remain in business,' Jim admits. Imported charcoal, consumer ignorance and centralized purchasing by the retailers are some of the problems. But the economic, environmental and moral arguments surely lie in his favour over the long haul. Demand for charcoal has risen greatly in Britain as people have become more enthusiastic about barbecues, but, Jim claims, more than ninety per cent of the charcoal supplied for this market comes from abroad, much of it as the by-product of uncontrolled timber extraction in the tropical forests of West Africa, South-east Asia and Brazil. Jim's timber is sustainably sourced – he has a coppicing lease from the Forestry Commission – but he says it would cost his small-scale operation too much to get the accreditation that would allow him to prove this to consumers by putting the Forest Stewardship Council symbol (seen on the copyright page of this book, for example) on his charcoal sacks. Meanwhile, British barbecue chefs unwittingly play their part in the razing of the Amazon, unaware that the very word 'Brazil' refers to burnt timber, the country having been named by the Portuguese after *brasa*, meaning 'hot coals', in reference to the red of the brazilwood trees that are being cleared at a rate of 10,000 square kilometres (four times the area of Dorset) every year.

Trying to survive in business has necessarily turned Jim into something of an environmental campaigner. But perhaps there is also something about the commodity he deals in that kindles the activist in him. For black carbon – charcoal or coal – has always been stuff for rebel causes. It is won by the poor to warm the rich. As long ago as 1662, John Evelyn gave a discourse to his fellow members of the Royal Society on trees and woodland culture called 'Sylva', in which he noted that all the charcoal made in the forests went for iron, for gunpowder and for 'London and the Court'. (Evelyn knew his business as his own family had a licence to make gunpowder for the crown.)

There is always a frisson between the getters and the consumers, the winners of the fuel and the winners in the end, that reminds us that energy is power. Coal miners' strikes are traditionally the bloodiest and most intractable of all industrial disputes. In *The Road to Wigan Pier*, George Orwell lionizes coal miners as the 'grimy caryatids' propping up the national economy. His famous description, both admiring and appalled, of a coal mine paints the men as 'splendid', the quantities hefted as 'monstrous', the noise as 'frightful', but the coal itself as merely black, an undifferentiated commodity to be attacked and demolished. In *Lady Chatterley's Lover*, D. H. Lawrence's Connie, Lady Chatterley, fears 'the industrial masses' and holds the miners in awe and dread; they are 'Fauna of the elements, carbon, iron, silicon . . . Elemental creatures, weird and distorted, of the mineral world!' Emile Zola's novel *Germinal* gives a graphic portrait of the collier's life in nineteenth-century France with a bitter strike at the heart of its story. After the defeated miners have returned to work, the eldest son of the principal family is killed in an underground explosion, and his body is brought up to the surface, reduced to 'black charcoal, calcined, unrecognizable'. We are what we mine.

Charcoal burners and the foresters for whom they often work excite similar fears and admiration, the more so since they at least appear to operate with a degree of autonomy, but also because the woods where they freely wander have always been the domain of outlaws. In the medieval period, the forests that covered much of Britain belonged to the king. 'Forest courts' decreed severe penalties ranging from death for taking the king's deer to blinding or castration for lesser offences. Even the taking of windfall timber was forbidden after the king usurped commoners' traditional rights and took control of ever more woodland for hunting. Charcoal burners needed a royal licence to burn timber for fuel and for use in forging iron. Charcoal-burning was thus one of the few more or less legitimate things one could claim to be doing in the forest if challenged by the king's men.

Tales of Robin Hood abound with disguises, including disguises as charcoal burners. A more authenticated medieval story concerns Fulk FitzWarin, a Shropshire gentleman sent in childhood to the court of Henry II, later dispossessed and driven to live as an outlaw. At court, he quarrels with the boy prince John. Later, when the outlawed Fulk learns that John, now king, is near by in Windsor Forest, he disguises himself as a charcoal burner in order to lure him deeper into the woods, saying that he has seen a magnificent stag. When he has the king at his mercy, he forces him to promise that he will restore his inheritance. John reigned at the beginning of the thirteenth century and decreed that the forest forges be shut down, perhaps after one too many encounters of this sort. Magna Carta, the English bill of rights that King John was forced to accept in 1215, was motivated in part by popular rejection of these draconian woodland powers.

The idea of strange men from the woods passing out gifts probably strikes us as creepy today, but it is a thread that runs from the myth of Robin Hood to the originally green-robed Santa Claus, who arises partly from the 'green man' of pagan religion. The association is not only with the trees but with their combustion products. In the Basque country, Santa Claus takes the form of the fat charcoal burner, Olentzero, who brings wooden toys that he has carved in his charcoal sack.

The redistribution of wealth and power was also one objective of the Carbonari, revolutionary precursors of the Risorgimento that would lead to the unification of Italy in 1871. They began as a secret society in the Kingdom of Naples, formed to resist French occupation during the Napoleonic Wars, taking their name from the Italian for charcoal burner, *carbonaro*. Their flag was red, blue and black for charcoal, only later becoming the red, white and green of modern Italy. The impulses of the Carbonari were patriotic, liberal and secular. After the defeat of Napoleon, they directed their efforts against their new overlords, the Austrians and the allied Papal States. The movement spread and, in 1820, after a number of failed uprisings, the Carbonari staged patriotic rebellions in several Italian cities. Shortly after eight in the evening of Friday 8 December 1820, the poet Lord Byron, then living in Ravenna, was caught up in one of these dramas when a powerful local Carbonari chief was assassinated. In *Don Juan*, he describes – 'This is a fact, and no poetic fable' – how he heard shots and ran out of his home to find the man lying in the street: 'for some reason, surely bad, / They had slain him with five slugs'. Though he distances himself from the crime – 'The man was gone: in some Italian quarrel' – Byron was himself active in the Carbonari movement, having been elected a capo, and was involved in buying and storing arms.

The Carbonari organized themselves along similar lines to the freemasons. The idea that they wore charcoal sacking and that their leader sat on a throne made of a charcoal bundle was merely an inspired piece of invented tradition to go with the romantic image of free men plotting liberty and independence in the forests of the Abruzzi. In reality, they were farmers and labourers, but also tailors, and even members of the junior clergy, who merely felt a certain solidarity with the sooty-faced practitioners of one of the most ancient crafts. The Italian Carbonaro was as ignorant of charcoal-making as the freemason is of stonework.

Carbon enjoys its economic centrality not because it is the only fuel, but because it is the only solid fuel with the convenient, and in fact essential, property that it burns away to nothing. In 1860, Michael Faraday devoted one of the Royal Institution Christmas lecture series that he had made famous to 'The Chemical History of a Candle' and explained to his youthful audience how the product of all carbon combustion is carbon dioxide, a gas that leaves no residue. Nearly fifty years before, he himself had seen this demonstrated in the most dramatic fashion in Florence by his mentor Humphry Davy, who burnt a diamond away to nothing using 'the great burning glass of the Grand Dukes of Tuscany'. In this behaviour, carbon is unlike almost all other combustible materials. If carbon left behind it the solid waste that metals leave when they burn – that is to say, an oxide heavier than the original material – the volume of waste from our hearths would be insupportable.

Gas though it is, even carbon dioxide has to go somewhere, of course. Faraday recognized this chemical quirk as an economic miracle, but he was not insensitive to what we would now call the carbon emissions of the Victorian city. 'A candle will burn some four, five, six, or seven hours. What, then, must be the daily amount of carbon going up into the air in the way of carbonic acid [carbon dioxide]!' A man in a day converts seven ounces of carbon from sugar in his body, Faraday calculated, and a horse seventy-nine ounces. 'As much as 5,000,000 pounds, or 548 tonnes, of carbonic acid is formed by respiration in London alone in twenty-four hours.' Faraday marvelled that plants were able to take up all this carbon dioxide, ignorant as he was of the level of the gas already building up in the earth's atmosphere. London's carbon emissions are today estimated as 44 million tonnes of carbon a year, 220 times the amount due to respiration alone in Victorian times.

## Plutonium Charades

Glenn Seaborg was arguably the greatest element discoverer of them all. He produced plutonium in 1940, curium and americium in 1944, berkelium and californium in 1949 and 1950 and had a hand in several others. His tally surpasses that of William

Ramsay, who discovered the inert gases, and beats the serial discoverers of new metals, Humphry Davy and, more significantly
perhaps, the great Jöns Jacob Berzelius of Stockholm.

For Seaborg, like so many discoverers of the elements, had
Swedish blood in his veins. His father's name was Americanized
from Sjöberg, his mother was Swedish, and Swedish was the
first language in the house where he grew up in Ishpeming in
northern Michigan, a region of the United States favoured by
Scandinavian immigrants who must immediately have felt at
home walking along earthen streets of packed iron ore.

Seaborg's high-school years had been punctuated by news of
chemists around the world excitedly claiming to have found the
last few elements that would plug the remaining gaps in Mendeleev's periodic table. The names they proposed invariably
declared a geographic allegiance: alabamine, russium, virgin-
ium, moldavium, illinium, florentium, nipponium. By the time
Seaborg was seventeen and graduating from school in 1929, the
periodic table seemed complete up to uranium, with ninety-
two protons in the nucleus of each atom, and therefore atomic
number ninety-two. Although some of these claims were erroneous or at least premature, it was eventually confirmed that the
elements we now know as technetium, astatine, promethium
and francium had been successfully synthesized in radiation
laboratories.

Seaborg was excited by the new realm on the border of physics and chemistry where chemical elements could be transformed
into one another and to which these powerful laboratories held
the key. As soon as he could, he was doing his own radiation
experiments. While still a graduate student at the University of
California at Berkeley, for example, he bombarded tellurium
with deuterium atoms and neutrons in order to convert it into a
heavy isotope of iodine whose radioactive presence could be
traced and used to monitor the functioning of the thyroid gland.

Tumours could then be found by using a Geiger counter to locate the hotspots where the iodine concentrated. Working with tellurium is always unpleasant – the compound that it forms with hydrogen is like hydrogen sulphide, with its infamous rotten-eggs smell, but far more offensive. Later, Seaborg managed to delegate the tellurium chemistry to his own student, who had great trouble ridding himself of the stink. Days afterwards, it was even possible to tell which library books he had been consulting from the revolting odour they exuded.

Seaborg was not content to leave his experiments in transmutation of the elements at that. He realized that the apparent ceiling on the number of the elements was only a matter of power. The strong nuclear force that binds neutrons and protons together to form the nuclei of atoms is only strong over extremely short distances. In larger atomic nuclei, the mutual repulsion of the positive electric charges of the protons becomes more important. 'At some point, the two forces could equalize. No one had realized that this might be why we had found no elements in nature with more protons than uranium's 92,' Seaborg wrote in a memoir.

The obvious thing then was to bombard uranium with particles and see whether any of them stuck. By early 1939 there were other reasons to do this. The world was arming rapidly in preparation for global war. Atomic fission had been reported by Otto Hahn in Nazi Berlin. Hahn had bombarded uranium atoms with neutrons and found not merely small particles breaking off as in a natural radioactive decay chain but whole atoms splitting in two – he was baffled to find barium, just over half the atomic mass of uranium, among his reaction products. His bafflement subsided somewhat when his long-time collaborator, the Jewish Lise Meitner (with whom he had discovered the element protactinium in 1918, and who was now in exile in Sweden), made calculations that confirmed the truth of what he had seen but not believed. She also noticed that heavy uranium, whose atoms contained more than the usual number of neutrons, could be expected to split into atoms of less massive elements with the release of huge amounts of energy. Seaborg's colleague Ed McMillan soon made similar observations, which led him to the conclusion that not all the uranium atoms split in this way, and that some might simply be absorbing the neutrons. If so, they would be transmuted into atoms of a new element, number ninety-three. This supposition was soon confirmed and the discovery published in 1940. By this time, Europe was at war, and the open publication of such potentially strategic information provoked a furious reaction from the British. It seems that the only thing that was kept quiet was the element's name: McMillan had chosen to call it neptunium, following the precedent of uranium, even though the planet Neptune had by then been known for nearly a century, but this information was not made public until after the war.

Seaborg's research into element number ninety-four was to proceed by contrast under a cloak of secrecy. Neptunium had too short a half-life for many applications, and certainly for making what was now being referred to as an 'atomic bomb' (although H. G. Wells seems to have coined the phrase in his 1913 novel *The World Set Free*). But there was reason to think the next element in the sequence would be different. The research began at Berkeley, but following the American entry into the war and the setting up of the Manhattan Project, the locus of the effort to synthesize plutonium moved to Chicago. Seaborg worked here for three years until 1945 in a building called, with deliberate obfuscation, the Metallurgical Laboratory, or Met Lab. The first task was to build an atomic pile in which slugs of uranium were stacked in such a way that they would undergo a chain reaction to produce element number ninety-four. At the start, the sought-for element was referred to simply as 94, but, as this was a little obvious, the chemists brightly adopted the code number 49 instead and took to calling it 'copper'. This was fine until one experiment actually required some copper, which then had to be distinguished by calling it 'honest-to-God copper'.

The new element was isolated in August 1942. Seaborg wrote – rather self-consciously – in his journal about 'the most thrilling day' in the Met Lab: 'Our microchemists isolated pure element 94 for the first time! It is the first time that element 94 (or any synthetic element, for that matter) has been beheld by the eye of man. I'm sure my feelings were akin to those of a new father who has been engrossed in the development of his off-spring since conception.'

Next, the offspring had to have a proper name. Extremium and ultimium were rejected, wisely in view of the chemical and military events that were to unfold. Seaborg instead followed McMillan's example and took advantage of the fact that there was one planet in the solar system left for inspiration, Pluto, which had been discovered in 1930. 'We briefly considered plu-tium, but plutonium seemed more euphonious,' he wrote later, insisting that the planets had been his only guide in choosing a suitable name. When he was reminded that Pluto is also the Roman god of the underworld and of the dead, Seaborg insisted that any such symbolic meaning was 'entirely coincidental; I was unfamiliar with the god or why the planet was named for him. We were simply following the planetary precedent.'

I think the chemist protests too much. Seaborg had literary leanings at school and came relatively late to science. It seems impossible he could have been unaware of Pluto's darker mean-ings. Certainly, his thinking was more knowing when it came to the chemical symbol. 'Each element has a one- or two-letter abbreviation. Following the standard rules, this symbol should be Pl, but we chose Pu instead,' he explained. P.U. – peee-euggh

– was and is American slang for a stink, something objection-
able. 'We thought our little joke might come under criticism,
but it was hardly noticed.' For certain key workers on the chemical side of the Manhattan Project, there was even 'the UPPU
club' – you pee plutonium. To qualify for membership you had
to have had enough exposure to plutonium for it to show up in
your urine.

Seaborg had his first microscopic speck of plutonium by
August 1943, a year after he had isolated the first invisible atoms.
Another year later, his reactors were producing masses of a
gramme or more, which were stockpiled at Los Alamos. With
the need to press on and complete the building of the bomb,
there was little time for meditation on the thrill of discovery,
and still less for much consideration of what plutonium was
actually like. In most cases, the discovery of an element is followed by a rush of chemists keen to measure its properties, test
its reactivity and prepare its compounds. In the case of plutonium, it was important to verify certain highly technical
parameters to do with its nuclear decay. But beyond that,
nobody seemed to care. Even the name – the usual sign of pride
in what one has brought into the world – had to wait before the
world could know it. At the end of the war, some of the Manhattan Project workers and their wives got together for a game
of charades, which confirmed that secrecy had been maintained:
'When the husbands tried to act out the word "plutonium," the
wives were mystified; they'd never heard of the stuff.'

The natural chemist in Seaborg reappeared much later. In a
1967 report called with perhaps unintentional poetry *The First
Weighing of Plutonium*, he described his new chemical element
with obvious awe: 'Plutonium is so unusual as to approach the
unbelievable. Under some conditions it can be nearly as hard
and brittle as glass; under others, as soft as plastic or lead. It will
burn and crumble quickly to powder when heated in air, or

slowly disintegrate when kept at room temperature . . . And it is fiendishly toxic, even in small amounts.' Despite all this, Seaborg fondly believed that plutonium might one day replace gold as a monetary standard. Maybe he really was oblivious of all plutonian symbolism.

The potency of plutonium was – and still is – felt in another sphere, of course. A few pounds of the element is enough for an atomic bomb, making it far more efficient than the alternative fissile isotopes of uranium. Werner Heisenberg and other German scientists were aware in 1941 that element number ninety-four might be a powerful nuclear explosive. However, it seems that the Allies never seriously entertained the possibility that the Nazis might be working on plutonium, while the Germans didn't realize that the Allies had it either. If either side had known of the other's interest and taken its implications into account in their military planning, the war might have run a very different course.

Plutonium, an element which hardly anybody has seen, has moved swiftly to occupy the demonic space traditionally reserved for sulphur, at first because of its use in the bomb and then because of gradually dawning public awareness of the difficulty of getting rid of it. The radioactive half-life of the isotope mainly present in plutonium nuclear waste is 24,000 years, which makes planning for its safe disposal an issue that transcends normal engineering considerations. Any storage structure must be sure to outlast the Pyramids and must communicate its deadly contents in a way that is sure to be understood by civilizations that will succeed our own.

As a budding chemist, I once applied for a summer job at what was then grandly called the Atomic Energy Research Establishment at Harwell in Oxfordshire. It was here that I had my first and only encounter with plutonium. The aura of power surrounding the element was made apparent when, as a condition of employment, I had to sign the Official Secrets Act. Was it the spartan accommodation they wanted to keep secret, or possibly the clapped-out military bus that ferried us to work? I passed the journeys knowingly reading *Catch-22* as the bus coughed its way along the weedy runways of the wartime airfield where the research establishment had pitched camp after 1945.

I found myself assigned to work in a laboratory led by a pipe-smoking figure with the windblown stride of Monsieur Hulot. The lab was designated 'red', the third of four levels of security. This meant I was cleared for laboratory work on dilute solutions containing plutonium and got to wear canvas overshoes which were good for gliding along the linoleum floors. Immediately, though, I felt a faint envy of those summer students who had been assigned to work in 'purple', the areas of highest security. The objective was to see how plutonium might be absorbed in material which could then be turned into blocks of glass. This vitrifying was thought to be a promising way to secure the waste for disposal by means and in locations never discussed. My experiment was always the same and involved pouring solutions of 'ploot' into columns containing the white titanium sand that was the raw material for the glass. I had no real sense of the dangers as I carried flasks of the radioactive liquid back and forth. It didn't glow green as it does in *The Simpsons*, nor did I find myself carelessly leaving work with test tubes of it stuffed in my pockets as Homer Simpson does at the Springfield reactor. (I don't recall ever being searched either.) My abiding memory is of the quiet tedium as the summer days slipped by while I transferred endless readings from the columns of sand into columns of figures on musty government stationery. It was the only time I worked in a laboratory.

Recalling those days, I feel a nostalgic urge to add plutonium to my own periodic table. I am missing all the natural elements with atomic numbers beyond eighty-two, lead; and of those above uranium which have to be manufactured artificially I have only Seaborg's americium, plundered from the mechanism of a domestic smoke detector where the stream of alpha particles emanating from it completes an electric circuit that is only broken if smoke blocks the path. I don't even have a piece of the highly collectible radioactive Fiesta chinaware made in the United States from the 1930s, whose papaya orange colour arises from uranium oxide used in its glaze.

Tracking down a specimen of the element I had once decanted in gushing quantities clearly isn't going to be easy. The reactors and research programme at Harwell were gradually wound down during the 1990s, amid accusations of contamination of the local water supply and, ironically enough, poor waste disposal practices. AEA Technology, the private company that inherited the business of the United Kingdom Atomic Energy Authority, has perhaps wisely changed tack and positions itself, slightly improbably, as a crusading consultancy on climate change. It is unable to help me. I try British Nuclear Fuels, the outfit in charge of Britain's nuclear waste, but find the telephone number of its director of corporate communications mysteriously cut off and later learn from its website that the company 'has progressively divested all its businesses and run down its corporate centre'.

The Americans seem more open about these things. Jeremy Bernstein's book *Plutonium* thoughtfully reproduces the specification of the isotope 239 of plutonium that is available for purchase from Oak Ridge National Laboratory in Tennessee. It is sold as oxide powder, at least ninety-nine per cent pure. 'This would be super weapons-grade plutonium.' There is a telephone number and an email address, isotopes@ornl.gov. I write requesting a small sample, plaintively adding that it would be a nice reminder of the hours I spent handling plutonium solutions as a student. The reply is as prompt as it is adamant: 'No we could not offer a sample of plutonium for any display.'

This seems a little mean-spirited. Plutonium appears to be restricted simply because, so far as its official guardians are concerned, the only conceivable reason anybody could have for wanting it is if they are planning to add to the global total of 23,000 nuclear warheads by building their own atomic bomb. The element's violent reputation is all that seems to matter; the fact that it is also a blameless occupant of the chemical pantheon, simply element number ninety-four, counts for nothing.

Besides, it is not as if I want a lot of it. The one course of action left to me is to pursue this logic to its ultimate conclusion. I learn that I can in fact easily buy 'plutonium' over the counter as a homoeopathic remedy. The point of homoeopathic remedies of course – incomprehensible to anybody with scientific training – is that they contain only the tiniest trace, or possibly even precisely none, of the stated active ingredient. So Plutonium (Homoeopathic Proving), a liquid distributed by Helios Homeopathy of Tunbridge Wells in Kent, presumably

contains an extreme dilution of some plutonium solution, perhaps of the kind I once worked with at Harwell. It seemed
perverse to name a product designed to appeal to soft-headed
mystics after the chemical element that has come to be seen as
the distillation of the human urge for self-destruction. The
Helios literature makes a wild stab at an explanation: 'The Pan-
dora's box of radioactivity has been opened and has released the
dark into the light,' it says. 'To rekindle the light our only
option is to enter this dark side fully. These radioactive materials, plutonium in particular, affect the deepest levels of the
human being – bone marrow, DNA, genetic structure, inner
organs, and the deepest emotions.'

I should say they do. Still, the fare to the dark side is a reasonable fourteen pounds. I dash to the Helios shop in Covent
Garden.

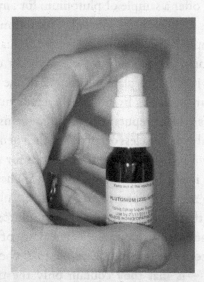

'I'd like some plutonium please,' I ask sweetly.

The assistant looks serious. 'I'll have to ask the pharmacist.'

The what? I wonder, looking up from reading the nonsensical

blurb on some remedy. There are muttered words from behind a wall of little brown bottles before the assistant returns. It seems the shop does not have any Plutonium. It's listed on the website, I point out helpfully. Reluctantly, the 'pharmacist' emerges from her lair and explains that they never have it – not that it's restricted or banned in any way, she adds. If I want to know more I will have to talk to head office. Then she breaks the shopkeeper's code of discretion and, through narrowed eyes, demands to know why I am interested in Plutonium anyway. I say I'm a chemist, and that I'd like some plutonium for my elements collection. Perhaps I should have said I wanted it in case I'm struck by some form of late-onset radiation sickness, but it's too late. She's exultant at having exposed an obvious homoeopathy sceptic.

At Tunbridge Wells, John Morgan is more helpful. 'There is no physical presence of the element,' he tells me. I suppose this is a homoeopath's idea of a guarantee. 'It is only the imprint of that element,' made by a process of 'molecular dilutions' or perhaps 'radionically', he's not sure. 'It's obviously impossible to go to a source material.' When it was 'proved', the remedy was judged to be particularly efficacious in dealing with depression. But, Morgan adds brightly, 'I suppose it could help to repair some damage if you had been exposed to plutonium.'

## Mendeleev's Suitcases

Blackballed by the Russian Academy of Sciences and overlooked in the first years of the Nobel Prizes, Dmitrii Mendeleev was only properly rewarded for his discovery of the periodic table nearly fifty years after his death. Then, finally, in 1955, honour was bestowed in the most fitting way, by naming after him one of the elements – the 101st – in the table. Astonishingly for this late date, Mendeleev was the first full-time chemist to be commemorated in this way. The elements that precede mendelevium in the periodic table, fermium and einsteinium, are named after physicists, reflecting their genesis in the great physics experiment known as the Manhattan Project. Later, other elements, too, would be named after physicists – Rutherford, Bohr and so on. The only elements celebrating chemists were gadolinium and curium, and even Marie Curie was as much a physicist as a chemist. It is the chemists' misfortune that the heyday of element discovery occurred in times more concerned to see honour done to nation and to Classical ideals than to their fellows. Today, their chance seems to have gone. It is unlikely now that we will see davium, berzelium, bunsenium or ramsayon.

Born in 1834, probably the fourteenth and last child of a Siberian family, young Dmitrii was taken by his mother to St Petersburg in the hopes that at least one of her children might improve himself. Like many aspiring scientists of the day, he travelled to Germany to complete his education on a government subsidy. His kind is unfairly satirized in several novels by Turgenev. However, for a Russian chemist of any ambition this was not dilettantism but an essential way to catch up on the latest developments in the science. Upon his return to St Petersburg in 1861, Mendeleev divided his time between the university, where he soon occupied the chair in chemistry, and expeditions to remote regions of the Urals and the Caucasus, where he acted as a consultant for the government and various commercial interests on everything from cheesemaking and agricultural productivity to the nascent oil industry.

The periodic table is one of those discoveries of science that suddenly explains so much that it seems it can only have sprung fully formed from the mind of its creator as if revealed in a dream. Mendeleev obligingly concocted a myth that this was exactly how he had come by it. But in the way of these things, his story of a dream came along rather late in the day. In fact, of course, the periodic table arose as the product of long cogitation. Mendeleev struggled to find a way of making sense of the elements for students as he worked on a much-needed introductory textbook in the Russian language. He wrote out the known elements with their atomic weights and some of their chemical characteristics on sixty-three cards. Then he began to group the cards as if playing a game of patience, placing the lightest elements in a row to begin with, but mindful that certain cards, for example those representing the halogens such as chlorine and iodine, seemed to belong together. He soon found that the lightest elements of each typical kind – the lightest halogen, the lightest alkali metal, and so on – provided a template for placing their heavier cousins. This breakthrough was made in the space of a day. From there, it might simply have been a matter of inserting all the remaining elements beneath the top-row element of their group in order of increasing atomic weight. But this is to reckon without the ambiguities among the sixty-three supposedly known elements, or the number of substances then tentatively accepted as elements that would later prove in fact to be some other element or combination of elements entirely. Both of these factors made it much harder for Mendeleev to be sure he had provided the best fit for the scientific evidence. The resulting 'Attempt at a System of the Elements, Based on Their Atomic Weight and Chemical Affinity' finally appeared in his textbook *Principles of Chemistry* in 1869, and only the following year, more confidently stated, in a scientific paper. He covered his bets by including variants of the layout that are today forgotten, and though by 1871 he was calling it 'periodic', it was to be many decades before all the cards were properly placed into their final, familiar pattern.

|        |            |             | Ti = 50 | Zr = 90      | ? = 180.   |
|--------|------------|-------------|---------|--------------|------------|
|        |            |             | V = 51  | Nb = 94      | Ta = 182.  |
|        |            |             | Cr = 52 | Mo = 96      | W = 186.   |
|        |            |             | Mn = 55 | Rh = 104,4   | Pt = 197,4 |
|        |            |             | Fe = 56 | Ru = 104,4   | Ir = 198.  |
|        |            | Ni =        | Co = 59 | Pl = 106,6   | Os = 199.  |
|        |            |             | Cu = 63,4 | Ag = 108   | Hg = 200.  |
| H = 1  |            |             | Zn = 65,2 | Cd = 112   |            |
|        | Be = 9,4   | Mg = 24     |         |              |            |
|        | B = 11     | Al = 27,4   | ? = 68  | Ur = 116     | Au = 197?  |
|        | C = 12     | Si = 28     | ? = 70  | Sn = 118     |            |
|        | N = 14     | P = 31      | As = 75 | Sb = 122     | Bi = 210?  |
|        | O = 16     | S = 32      | Se = 79,4 | Te = 128?  |            |
|        | F = 19     | Cl = 35,5   | Br = 80 | J = 127      |            |
| Li = 7 | Na = 23    | K = 39      | Rb = 85,4 | Cs = 133   | Tl = 204.  |
|        |            | Ca = 40     | Sr = 87,6 | Ba = 137   | Pb = 207.  |
|        |            | ? = 45      | Ce = 92 |              |            |
|        |            | ? Er = 56   | La = 94 |              |            |
|        |            | ? Yt = 60   | Di = 95 |              |            |
|        |            | ? In = 75,6 | Th = 118? |            |            |

The difficulty for everybody else was that Mendeleev's table seemed to come from nowhere. For several years there was no telling whether it was true or false. What could be 'true' anyway about an arrangement of symbols on paper? The Russian claimed his table could be used to predict important properties of the elements such as densities and melting points, but the fact that it did this from an entirely theoretical standpoint was merely grist to the mill for his opponents.

However, the critics were silenced in 1875 when Paul-Emile Lecoq de Boisbaudran, entirely unaware of Mendeleev's work, announced that he had discovered a new aluminium-like element, which he named gallium. Its atomic weight corresponded exactly to the value Mendeleev had assigned to a gap in his table directly below aluminium, and even the mode of its discovery – by identification of its characteristic spectrum – was as he had predicted. Lecoq reported a density rather lower than the Russian had estimated, but Mendeleev brazenly wrote to Lecoq suggesting he prepare a purer sample. When he did, the density very closely matched Mendeleev's value, dramatically vindicating the Russian's theoretical science. (Gallium's most striking property, however, its low melting point, had been anticipated by no one – it melts in the hand, making it only the second metal, after mercury, readily observed in the liquid state.)

The story was repeated in 1879, when Lars Nilson at Uppsala University filled the space Mendeleev had left between calcium and titanium with the discovery of scandium, and again in 1886, when Clemens Winkler at the Freiberg mining university in the ore-laden mountains on the border of Saxony and Bohemia isolated the semi-metal germanium, intermediate in the periodic table between silicon and tin, from a local mineral specimen.

Subsequent printings of Mendeleev's *Principles* filled in each gap as the news came through, and the 1889 edition went so far as to print photographic portraits of Lecoq, Nilson and Winkler, lionizing them as 'reinforcers of the periodic law'. Though now honoured by many foreign science academies, Mendeleev was nevertheless blocked from higher recognition at the St Petersburg Academy of Sciences because of concerns about his anti-imperial politics, the seeds of which had been sown in his Siberian youth when he fell in with a group of exiled Decembrists, the failed revolutionaries who tried to topple Tsar Nicholas I in 1825. Later he was forced to resign his professorship at the university. Ironically enough, he quickly found alternative employment in advisory positions to the government.

For a while, each discovery of a new element drew an appreciative response from Mendeleev when it fitted into his grand plan. But in time, more sophisticated techniques came along that were able to reveal new elements with unforeseen properties that could not be so readily embraced. William Ramsay's discovery of the inert gases, beginning with argon in 1894, was the first major interrogation of the periodic table after twenty-five years of successful consolidation. Mendeleev had once again observed that there were gaps, based on atomic weights, between the alkali metals and the halogens, but on this occasion the scarcely believable implication was that an entire family of elements was missing, and it was less clear how, or indeed whether, the table should be amended. The 1895 edition of his still standard textbook entered a note of scepticism about the first reports of argon and helium. There followed a tetchy correspondence between the two men, with Mendeleev at first refuting Ramsay's claim and suggesting that his new gas argon was simply a heavy form of nitrogen. (Like the ozone form of oxygen, which contains three atoms rather than two, this putative three-atom molecule would be half as heavy again as the normal nitrogen molecule of two atoms, bringing it close to the observed weight of Ramsay's argon.) As Ramsay added more elements of similar character, first helium, and then neon, krypton and xenon, in rapid succession, Mendeleev came round to the idea that they could after all be accommodated in his system by the simple expedient of adding a new column to the edge of the table. Astonishingly, it appears that Mendeleev's failure to predict the inert gases after so many other successes may have been a major reason why the Nobel Committee decided not to award him the prize in chemistry when they considered the possibility in 1906.

The discovery of the radioactive decay of elements by Marie Curie and others in Mendeleev's declining years played further havoc with his system of chemical order. What was the point of putting elements in boxes if they could simply jump from one box to another by shedding a few subatomic particles? Mendeleev had once taken to the road in Russia in order to do battle against the spiritualism that he felt was preventing progress in the country; visiting the Curies' laboratory in 1902, he felt he was once again dealing with the same ungovernable forces, which he scathingly called 'spirit in matter'.

Mendeleev has frequently been characterized as a mystic and

a prophet, but this is more to do with his Siberian origins, his irascibility and his dishevelled beard than his professional record. Contemporary portraits don't always help: one shows the chemist leaning back in his chair maniacally clutching a book to his face with both hands, a glowing cigarette held in his fingers. Mendeleev had brilliantly devised a periodic system of the elements in which he had sufficient confidence to leave gaps, but this was a sensible conjecture based on scientific evidence, not prophecy. His other activities were equally grounded in rationalism – tackling spiritualism, advising on the national economy, recommending agricultural reforms. Though full of ideas, he was by nature something of a conservative and, while not accepted into institutions such as the Academy of Sciences, he still seemed like an establishment figure to others. The final seal of conventionality surely came in 1893, when he was put in charge of the newly founded national board of weights and measures.

Shortly before becoming a professor, Mendeleev had bought a summer estate outside Moscow. Like Levin in *Anna Karenina*, he used the land to showcase his ideas of progressive farming. Here, his daughter Liubov' Dmitrievna Mendeleeva met and fell in love with the young poet Alexander Blok, whose family owned a neighbouring estate. In 1903, the year of their marriage, Blok wrote admiringly to Mendeleeva of her father, who 'knows everything that happens in the world already for a long time. He has entered into everything. Nothing is concealed from him. His knowledge is most complete.' Blok – the author of 'The Scythians' and other works in which a Russian identity rooted in the wildest regions is given voice in the language of the literary avant-garde – surely responded to Mendeleev's incongruous blend of deep Russian ancestry and immersion in the latest currents of thought in scientific Europe. After Mendeleev's death in 1907, Blok contrasted him favourably with the cynical establishment intelligentsia for holding to an optimistic view of the country's future. But later something snapped, and the poet, filled with revolutionary zeal, decided that his father-in-law belonged too much to the past. On 31 January 1919, he wrote in his diary: 'Symbolic action: on the Soviet New Year I smashed Mendeleev's desk.'

Mendeleev's university apartment – though sadly not the laboratory that once adjoined it – is now preserved as a museum. I visited it one blistering June day, crossing the glittering Neva in a dazzle of golden domes to find myself strolling along the elegant terraced avenues of the university complex on the grid-planned Vasilevskii Island. The entire place still glowed with the sense of Peter the Great's ambition to found a city that would rival the greatest in Europe.

This was where Mendeleev lived for twenty-four years, from his appointment to the chair of chemistry in 1867, through the time when he worked out the periodic table and enjoyed the satisfaction of seeing his predictions of missing elements realized, to his forced retirement in 1890. The rooms were crowded with heavy armchairs and sofas and equally heavy volumes of journals. In one, a cigar-smoking portrait presided over the scene. Photographs of Mendeleev with scientists, including the discoverers of his predicted elements, and leading figures in St Petersburg lined the walls. The signatures of his visitors were illegibly inscribed on a tablecloth. There was also a desk. Was this where he had laid out his element cards, or was that on the desk that Blok demolished? The pack of cards and other documents showing Mendeleev's workings are long lost, but his textbook survives, and so does the periodic table in it, the sequence of the elements instantly recognizable, even though the whole thing is twisted through ninety degrees, making rows into columns and columns into rows. Thus, B, C, N, O, F appears as a column on the left; Al, Si, P, S, Cl to its right. As the atomic weights increased, I noticed alignments that we would now think of as misleading – mercury grouped with copper and silver, for example, while gold was aligned with aluminium. But there were also the question marks against gaps in the sequence that were the true sign of Mendeleev's genius.

Seeing the familiar array of letters in cold print, it was hard to believe it had not swept all before it. I asked the museum's curator, Igor Dmitriev, why this was. 'There were many classifications already,' he explained, 'none of them taken seriously. So it is understandable that Mendeleev would have had a hard time.'

But it was the suitcases that really stuck in my mind. Mendeleev may not have been a mystic, but he certainly had his eccentricities, and one of the oddest of them was his hobby of making leather suitcases. His apartment was cluttered with cases in varying states of completion, as well as the leather and buckles and tools used to make them. It's tempting of course to see this curious pastime as a metaphor, as material evidence of the character of a man obsessed with packing things neatly away. But it's neither necessary nor helpful to do this. In truth, Mendeleev could confess to his fair share of nineteenth-century science's passion for organizing nature – he had been attentive to contemporary naturalists' efforts to classify living species, for example. But his system for the chemical elements, the ultimate pigeonholing of nature, sprang simply from a pedagogical need to streamline the presentation of chemical knowledge rather than from rage at the disorder of the world.

Mendelevium was the first element that had to be dragged into the world atom by atom, beginning in 1955. Even now, it has never been made in quantities visible to the eye. 'We thought it fitting that there be an element named for the Russian chemist Dmitrii Mendeleev, who had developed the periodic table,' wrote its discoverer, Glenn Seaborg. 'In nearly all our experiments discovering transuranium elements, we'd depended on his method of predicting chemical properties based on the element's position in the table.' At the height of the Cold War, this 'somewhat bold gesture', as Seaborg admits, was condemned by some Americans, but was not unappreciated in top Soviet circles. The tiny amounts of mendelevium that have been made in the particle accelerators at Berkeley and elsewhere decay rapidly, and it has not been possible to make more than a start on measuring its essential properties or investigating its chemistry. One suspects that this would have bothered Dmitrii Mendeleev, the supreme theoretical chemist of his day, not one bit.

## The Liquid Mirror

In Jean Cocteau's 1949 film *Orphée*, Orpheus enters the underworld in pursuit of Eurydice by passing through a mirror of mercury. The scene is a masterly cinematic sleight of hand. Orpheus, played by a Grecian-coiffed Jean Marais, is led to a large dressing mirror. He dons latex gloves – a magical preparatory ritual that doesn't entirely disguise the fact that Cocteau, the renowned avant-garde artist, seems to have had a thoroughly modern concern for health and safety. 'With these gloves, you will be able to pass through the mirror like water', Orpheus' guide explains. 'First the hands.' Doubtfully, Orpheus does as he is told and puts his palms to the reflective surface, and is met by its resistance – it's just a mirror. 'Il s'agit de croire', he is advised: you must believe. Then we see his fingers in close-up pushing through the barrier, its surface set aquiver by the fateful action. The film cuts to an overhead shot. With the liquid mirror surface now hidden from our view, Orpheus and his guide disappear through the portal.

It is axiomatic that we cannot know the underworld until we ourselves leave the world, and for this reason Cocteau sought for his divide between the two a total optical barrier that was nevertheless physically penetrable. The set-up is said to have required a reservoir of half a tonne of mercury. This seems excessive until one remembers that this metal is so dense that lead will float on its surface. A pool of this weight the size of a full-length mirror would be not much more than a centimetre deep. It is of course not possible to arrange for such a pool to stand upright, so Cocteau had to turn his camera to produce the illusion of a vertical mirror for the brief scene where Orpheus' hands pass through the barrier. And it is not possible, or safe, for whole bodies to be immersed in mercury, hence the subsequent cutaway to overhead.

The artist might have used milk or paint to achieve some of the necessary effect, but mercury was well chosen as the only liquid able to provide a perfect reflection. The material also offered a serendipitous bonus. In an interview, Cocteau later explained: 'In mercury the hands disappear, and the gesture is accompanied by a kind of shiver, whereas water would have produced ripples and circles of waves. On top of that, mercury has resistance.' In this single action, then, are made visible signs of Orpheus' trepidation, of his fright and of the effort of will he must summon in order to abandon life. Furthermore, the unfamiliar, almost *un*natural, quality of the mercury hints nicely at uncertainties to come in the *super*natural world.

Known for perhaps 5,000 years, mercury has always been celebrated for its unique confluence of liquid and metallic properties, even if this made it no easier for people to find a use for the stuff. For a material that is clearly special, yet also rather useless, there is one obvious application, and that is in sacred rites. Cocteau's employment of mercury as the gateway to another world is merely a modern twist in a long and universal tale.

The first emperor of China, Qin Shi Huang, who unified the country in 221 BCE, is said by legend to lie buried beneath a rugged verdant mound near Xi'an in the Shaanxi Province of northern China. The historian Sima Qian, writing a century after the emperor's death, describes a vast bronze-lined chamber, its ceiling jewelled to represent the heavens, containing a fantastic model of the emperor's palace, his capital city Xianyang lying around it, and his entire empire beyond. Through the model landscape are said to run channels of mercury representing the hundred great rivers of China. Although it is not easy to see how it could be done, Sima writes of mechanisms to pump the heavy liquid round, maintaining a continuous flow that symbolizes the eternal lifeblood of the emperor. It is likely, what's more, that Qin's blood actually did contain mercury at the time of his death as he is thought to have swallowed mercury pills in the hope of obtaining immortality.

It was in this region of China in 1974 that archaeologists began to uncover the now famous Terracotta Army, hundreds of life-size earthenware figures, soldiers first, then later musicians, athletes and bureaucrats, providing extraordinary details of life at the beginning of the Qin Dynasty. The location of the find was soon matched with descriptions of the landscape in Sima's history, and from this it was surmised that a particular eminence a kilometre off to the west might hide the emperor's tomb. Subsequent excavations have revealed that the pits containing the Terracotta Army were just part of a large underground complex around this feature, but the mound itself has not yet been broached for fear that it may not be possible to preserve its contents – not least its fabulous mercury rivers – if they are disturbed. However, scientists have carried out various non-destructive tests at the site, including chemical analysis of soil samples. These have revealed levels of mercury well above the normal in the immediate vicinity of the burial mound. In Sima's account, the model empire is carefully oriented underground to correspond with the real geography, and it has been found that some of highest concentrations of mercury align with some of China's coastal seas and the vast sweep of the lower Yangtze River.

The Chinese obtained mercury metal readily enough from the abundant red ore cinnabar, and this pigment has itself permeated the culture in the form of the ubiquitous vermilion that is regarded as a uniquely auspicious colour. Cinnabar was strewn in graves to restore colour to the cheeks of the dead, and as early as the Shang Dynasty, 1600 BCE, it was being used to make ink with which to tint the Chinese characters incised in pieces of bone. The metal itself was used as an alternative liquid to drive water clocks or in mechanized armillary spheres. It was even used to make tumbling toys. 'The Chinese have probably used mercury and cinnabar more extensively than any other people,' according to the great sinologist Joseph Needham in his twenty-four-volume *Science and Civilisation in China*.

A modern mercury cascade with its own message of life and death was created by Alexander Calder for the Spanish Pavilion at the 1937 Paris Exposition. The American artist received the commission indirectly from the short-lived republican government in the midst of Spain's civil war, and his *Mercury Fountain* duly went on display in the same space as the documentary masterpiece of those years, Picasso's *Guernica*. Calder's work is more oblique in its reference to the conflict. The mobile sculpture consists of a series of three metal plates arranged above a large pool of mercury. Mercury is pumped up so that a fine stream trickles on to the top plate. It quickens in droplets and rivulets

across the plates in turn while they gyre and bow under the weight of the metal, before it vanishes quietly into the pool below. The mercury is the key to the meaning of the work. It came, like the majority of the world's mercury at that time, from the cinnabar deposits at Almadén in Ciudad Real south-west of Madrid. This strategically important location was to be repeatedly besieged by Franco's insurgents, and Calder's work commemorates the miners who had successfully held off the first nationalist onslaught a few months earlier. In one of the most imaginative war memorials ever devised, we see bright lives aggregating, separating, shaping larger events, and those events in turn determining their fate, before their ultimate absorption into stillness.

Almadén means 'the mine' in Arabic, and this site was well known to the Arabs who ruled Spain from the eighth to the fifteenth centuries. Calder's fountain is an acknowledgement of this history too. In 936, at Medina Azahara near Córdoba, some hundred kilometres to the south of Almadén, the caliph Abd al-Rahman III began to erect a vast personal estate with a mosque and gardens overlooked by a sumptuous palace. An enchanting feature of this richly decorated alcázar, or palace complex, was a pool of mercury positioned such that it bounced bright shafts of sunlight round the interior of the room in which it was situated. Guests were able to dabble their fingers in the metal, enjoying its cool, enveloping touch, and sending wild dapples across the ceiling like an early version of a dance-floor glitter ball. Ornamental pools of mercury were a feature of Islamic high living, and there is evidence that they were also used in pre-Columbian America. Before the element's poison-ous qualities were known, it was natural in places where it was easily obtained to rejoice in the coursing, trickling, glittering quality of the liquid.

When it was moved to the Joan Miró Foundation in Barcelona

in 1975, *Mercury Fountain* was put on display in its own glass cubicle. No longer could visitors do as they had done in Paris and throw coins on to the liquid surface just to see them float there. It had in fact shown a remarkably lax attitude to public health that the viewers were allowed to enjoy such open access to it in 1937. Of the 200 litres of mercury that arrived from Almadén on the afternoon of the press opening of the Spanish Pavilion (Calder had used steel ball bearings to mock up the action of the sculpture as he worked on it), an astonishing fifty litres were to be held in reserve, said Calder, to take care of losses due to splashing and leakage over the course of the exhibition. The toxic effects of mercury – familiar as an occupational hazard of hatters and others who used mercury compounds in their work – are felt when it is absorbed through the skin or when vapour enters the lungs. Yet for admirers of Calder's artwork there was not even Cocteau's rudimentary precaution of latex gloves.

The quarantining of *Mercury Fountain* is emblematic of what is happening to the element everywhere. From its beginnings as a decorative and mystical wonder, mercury went on to find many uses exploiting its exceptional combination of properties – density, fluidity, conductivity. Its compounds have been used as pigments and cosmetics. Their often poisonous nature suits them for insecticides and marine antifouling. In medicine, they have provided the active ingredients of everything from drastic syphilis treatments to routine laxatives and antiseptics such as calomel and Mercurochrome. But all these and many other applications are now falling into disfavour. On 1 January 2008, Norway banned all imports and manufacturing involving mercury, even including the making of dental amalgams. The European Union is to ban the export of mercury from July 2011 in an effort to reduce global exposure to the element. Mercury thermometers and barometers will become historical relics.

Almadén has finally ceased production after more than 2,000 years of operation. With mercury stopped at its source, attention turns to what is already in circulation. A British study of cremations has even raised concern about the element escaping into the environment as the fillings in the teeth of the deceased are vaporized – the spectre of our once-easy coexistence with the metal come to haunt us.

Soon only highly specialized applications may be left, although it is some consolation that one or two of these recapture the surreal delight of more ancient mercury amusements. In the mountains of British Columbia not far from Vancouver is the Large Zenith Telescope, which obtains its images of the heavens using a liquid mirror. Mercury is poured into a wok-like dish six metres in diameter. The dish revolves at a stately pace, forcing the surface of the mercury into a paraboloid more perfect than can be obtained by solid glass or aluminium. The idea had been around for more than a century, but it is only recently, as the metal was drawing opprobrium elsewhere, that it has become possible to create a sufficiently smooth-running mechanism to enable sharp images to be produced from such a mercury pool. Liquid-mirror devices must of course be held horizontal if they are not to spill their magical fluid. Constrained to gaze ever upward, these telescopes do not scatter sunlight but gather the light of the stars, offering a window not on the underworld but out on to other worlds.

Many chemical procedures that were well known to the alchemists now lie beyond the bounds of normal scientific practice, not because they are especially complicated or obscure, but because they are regarded as so hazardous that modern health and safety laws will not permit them to be undertaken even with all the safeguards of a state-of-the-art laboratory. One of these procedures is the reversible combination of mercury and sulphur, a reaction that was central to alchemical theory. The alchemists' interest in this simple reaction is easily explained. By putting yellow sulphur, which is dry and hot, together with liquid mercury, which feels cool and wet, they brought together the four principles of all matter.

The colour of the sulphur and the bright gleam of the mercury suggested furthermore that gold might be the outcome of the fusion. The alchemists believed that all metal deposits in the earth were on their way to becoming gold; if a man found instead tin or lead he had simply come too early. With their auspicious appearance, mercury and sulphur, both frequently occurring in their native state, looked to offer a faster route to this goal. The great Arab alchemist and mystic of the eighth century, Jabir ibn Hayyan (his name often appears latinized as Geber), who may have been responsible for bringing Chinese knowledge of cinnabar and mercury to the West, believed that perfection in metals, whether found in nature or made by man, could only be achieved when these two elements were present in the correct proportion and at the right temperature. Lack of perfection – that is to say finding base metal when one hoped for gold – was simply explained as a disproportion of these factors. In Jabir's view, the more precious metals were made by ensuring that a relatively greater amount of mercury was present. But there were further provisos to do with the purity and type of each element used. Silver was to be made by combining mercury with what Jabir called white sulphur, for example, whereas gold was made from 'best' mercury with only a little red sulphur, though it is impossible to be sure exactly what he meant by these terms.

That was the theory. Experiments proved disappointing, needless to say, although some disreputable practitioners managed to persuade a few credulous souls that they had at least increased the quantity of their existing gold through the addition of mercury and sulphur – the sulphur would have burnt off while the mercury merged with the gold by amalgamation, producing an apparent gain in weight, but of course no more gold. Rather than abandon their cherished hope, the alchemists elaborated Jabir's theory in the light of these unsatisfactory results by suggesting that all manner of metals in addition to gold might be brought forth simply by juggling the relative proportions of these two elements. This reaction was therefore at the heart of mainstream science in medieval Europe, and remained core to alchemical thinking for several centuries. It was often performed and it enjoyed scholarly approval. One early seventeenth-century text shows an engraving of Thomas Aquinas pointing like a holiday tour guide towards an instructively

AVREAE MENSAE, LIB. VIII.    365

EI SVLPHVRE ET ARGENTO VIVO,
*ut natura, sic ars producit metalla.*

HOMAS AQVINAS ob singularem pietatem & deuotionem *Sanctis* adnumeratus in doctrinis & scientijs tam admirabilis toti Christiano orbi apparuit, vt nomen *Angelici doctoris* adsciuerit; tanquam supra humani ingenij vim & captum ad spiritualem naturam ascenderet: Qui sane titulus si cui hominum conuenit, illi imprimis inuidendus non erit. Tanta enim ille opera quaestionibus subtilissimis & in diuinis & humanis

*Thomas Angelicus d. stor.*

Zz 3                 refer-

cutaway turf-covered furnace in which the vapours of two elements intertwine. 'As nature produces metals from sulphur and mercury, so does art,' reads the caption. This reaction, though undertaken on the basis of erroneous belief, was nevertheless a key turning point on the road to modern chemistry. It was perhaps the first instance of the informed synthesis of a new substance from two known ingredients. Furthermore, it was the first clear demonstration of the reversibility of chemical reactions – for not only did mercury combine easily with sulphur to form mercury sulphide (cinnabar), but the mercury sulphide also, when heated, separated back into its two constituent elements – so providing an important hint that matter could be neither created nor destroyed.

It is not a difficult experiment. I could easily cannibalize the mercury from an old thermometer, put it in a crucible, mix in an appropriate amount of sulphur, cover it, and heat it until the rich vermilion colour of mercuric sulphide began to appear. I could heat it again in order to separate these two constituent elements, distilling off the mercury as the sulphur burnt away. But, while I am sceptical of the advertised hazards of the many chemical experiments one is these days discouraged from attempting at home, I am aware now (as I wasn't when I used to obtain my mercury by roasting dead batteries) that mercury vapour is deeply unpleasant stuff.

I settle for witnessing the experiment at one remove with the assistance of Marcos Martinón-Torres at University College London. Marcos has carved out an academic career at the junction of archaeology and materials science that gives him a marvellous pretext for re-enacting the experiments of the alchemists in the interests of historical accuracy. When it comes to repeating the mercury–sulphur experiment, however, even he is banished from the laboratories of his institution and obliged to make himself scarce in a secret field hidden in the suburbs.

The reaction vessel is a clay aludel – an Arabic word, like so many in chemistry – which is a kind of large crucible with a high pointed lid like a witch's hat where vapours can mingle and cool. The contraption is about the size and shape of an ostrich egg. A small vent at the top prevents the pressure from building up inside the vessel and causing an explosion. Marcos and a colleague, Nicolas Thomas from the Panthéon-Sorbonne University in Paris, sprinkle the cinnabar they have brought into the bottom of the aludel, put the hat on and seal it shut with wet clay. Then they build a small furnace of bricks and clay and fill it with charcoal and light it. When they judge it is hot enough to decompose the cinnabar, but not so hot that the mercury will escape as vapour, they place the aludel in the furnace. Wearing breathing apparatus, they crouch at the field edge, watching the aludel carefully as the red heat of the fire begins to warm it. Relieved that it has not cracked, they soon observe small beads of mercury that have condensed around the vent hole. This is the sign that the reaction has occurred. They allow the vessel to cool and then break it open. A firmament of tiny bright globules

has settled over the inner wall. By collecting the mercury, adding sulphur and heating once more, they recover the cinnabar, a yellow and orange mess, part solid, part melt, looking for all the world like a steamed treacle pudding but smelling of the devil.

# Part Two: Fire

## *The Circumnavigation of the* Sulphur

Gold and silver, iron and copper all appear scores of times in the Bible owing to their monetary or utilitarian value. Lead and tin are mentioned in passing. These are six of the ten elements known from antiquity. One further element has a symbolic value of an entirely different kind, and that is sulphur, or brimstone as it is universally termed in English Bible translations.

Brimstone earns fourteen mentions in all, and not one of them is complimentary. Its every appearance is accompanied by scenes of punishment and destruction or at least the threat of great violence. In Genesis, the depraved cities of Sodom and Gomorrah are overthrown when 'brimstone and fire from the Lord out of heaven' is rained upon them. Six of the references come in the central chapters of the book of the Revelation of Jesus Christ given to the Apostle John, and cover the Great Tribulation, the Return of the King, the Millennium and the Last Judgement. The sulphur starts to flow once the seven seals have been opened and six of the seven trumpets have sounded, and scarcely lets up until the New Jerusalem is revealed 200 verses later.

In chapter nine of Revelation, John sees a third of humanity slain by an army of 'two hundred thousand thousand' horsemen. The riders have

breastplates of fire, and of jacinth, and brimstone: and the heads of the horses were as the heads of lions; and out of their mouths issued fire and smoke and brimstone. By these three was the third part of men killed, by the fire, and by the smoke, and by the brimstone, which issued out of their mouths.

Then the seventh trumpet sounds, proclaiming God's kingdom in heaven. Satanic beasts rear their many heads, and an angel warns that anybody who worships the beast 'shall drink of the wine of the wrath of God, which is poured out without mixture into the cup of his indignation; and he shall be tormented with fire and brimstone in the presence of the holy angels, and in the presence of the Lamb'.

Babylon falls, heaven rejoices, and Christ appears on a white horse. In the Battle of Armageddon that follows, the Devil and his accomplices are 'cast alive into a lake of fire burning with brimstone'. Finally, John hears God pronounce sentence on the remainder of people who reject His word: 'the fearful, and unbelieving, and the abominable, and murderers, and whoremongers, and sorcerers, and idolaters, and all liars, shall have their part in the lake which burneth with fire and brimstone: which is the second death'.

God, or John, shows so little imagination in the forms of punishment administered during the last days that we must take it that fire and brimstone have a particular ritual significance. The fact that hellfire is always accompanied by brimstone, that brimstone is never present without fire, indicates not only that brimstone is flammable but also that there is something specially horrific about its flame. Milton was well aware of these properties, which are crucial in setting the opening scene of *Paradise Lost*, where we find the Devil cast out from heaven in

> A Dungeon horrible, on all sides round
> As one great Furnace flam'd, yet from those flames
> No light, but rather darkness visible

> Serv'd onely to discover sights of woe,
> Regions of sorrow, doleful shades, where peace
> And rest can never dwell, hope never comes
> That comes to all; but torture without end
> Still urges, and a fiery Deluge, fed
> With ever-burning Sulphur unconsum'd.

For sulphur burns not like a candle, but with a low blue flame that is barely luminous – 'darkness visible' indeed. It is not rapidly consumed like a wood fire, so that it is easy to imagine the flame as 'ever-burning', especially if the ignited sulphur occurs, as it sometimes does in nature, in a seam that runs endlessly and invisibly into the earth.

Could this dreadful material really be the same as the sulphur I'd once seen stacked up on the docks at Galveston in Texas? Lemon-yellow bricks of the stuff, each the size of a container truck, were arrayed several high and several deep, the cheerful colour making them look more like an unusually successful piece of public art than a vital industrial commodity awaiting shipment. The substance was the form of the element purified by sublimation – that is, by condensing the solid directly from the vapour – known quaintly as flowers of sulphur, and in the spring sunshine ideas of hellfire and damnation were very far away.

Elemental sulphur is bland enough; its disagreeable alter ego is only awakened when it undergoes chemical change. The simplest reaction is combustion, which produces the corrosive, bleaching and choking gas sulphur dioxide. Its effect is cleansing as well as burning – something that obliges us to begin to differentiate between simple fire, which is destructive, and biblical brimstone, whose burning stench can also be purgative: perhaps through the action of sulphur even Satan might be redeemed in his former guise as Lucifer, the angel who fell from heaven. In antiquity, sulphur was widely used as a disinfectant and for related ritual purposes. When Odysseus returns to Ithaca and slaughters the suitors who have been pestering his wife, Penelope, he orders the nurse to 'bring some sulphur to clean the pollution, and make a fire so that I can purify the house'. Sulphur is still sold for this purpose today, the suggestion being that you use it in the greenhouse rather than to dispel unwanted personal attention. Sulphur fires were used to combat cholera into the twentieth century, and sulphur was taken internally for digestive and other complaints. Mrs Squeers holds 'brimstone-and-treacle mornings' at Dotheboys Hall in Dickens's *Nicholas Nickleby*, the filthy mixture administered, as she explains, 'partly because if they hadn't something or other in the way of medicine they'd be always ailing and giving a world of trouble, and partly because it spoils their appetites and comes cheaper than breakfast and dinner'.

Combustion is a rapid form of oxidation, the chemical combination of a substance with oxygen, while what goes on in the stomach is the opposite process, known as reduction, which is accomplished by the action of bacteria. The simplest reduction of sulphur yields another foul gas, hydrogen sulphide. Between them these two basic chemical processes account for a vast range of sulphur chemistry that is essential to life. The handful of noxious compounds so made undoubtedly give the element its evil reputation, but this reputation would not exist if these compounds were not locked into a cycle with others responsible for more pleasant sensations. For example, the various pungent odours of the alliums arise from this chemistry, with onion, garlic, leeks and chives each containing a different sulphur compound in minute quantities. During cooking, these compounds are converted into substances far sweeter than sugar, related to those used in artificial sweeteners. In the cabbage family, on the other hand, cooking gradually converts sulphur-containing compounds into smellier forms, which is one of the things that makes overcooked brussels sprouts so unappealing. The sulphur compounds released when we digest our food are passed out of the body in excrement, and especially, because so many of them are volatile, as flatus and bad breath. One of them, methyl mercaptan, allegedly the world's smelliest molecule, is added to otherwise odourless natural gas so that we may be sensitive to leaking pipes. Though the sulphur is only present in tiny amounts, its foul odours and association with bodily functions are enough to explain its devilish cultural reputation among the elements.

The sulphur I saw on the quay at Galveston was a by-product of the local petrochemical industry. It made me think of the fumaroles under the Gulf of Mexico where specialized marine bacteria synthesize pure yellow sulphur out of the gases released from the bowels (the cliché apt for once) of the earth. I knew, of course, that the element was in fact recovered from hydrogen sulphide in the natural gas brought ashore from the offshore platforms, but in both cases the gas is ultimately the product of the decay of palaeozoic plants. Even the 'smell of the sea', it has recently been found, is owing to a sulphurous gas, this time dimethyl sulphide, released by living microbes in surface waters.

The smell of the sea must have beckoned the sailors who embarked at Plymouth on Christmas Eve 1835 at the beginning of what was to be a seven-year circumnavigation of the globe with the purpose of surveying the oceans and collecting scientific specimens. Their ship was called HMS *Sulphur*.

The expedition was similar in intent to that of HMS *Beagle*, which was then on the final leg of its own long voyage, and shortly to set ashore its dangerous cargo of Charles Darwin and all his specimens and new ideas. It is recounted in the two volumes of a *Narrative of a Voyage Round the World, performed in Her Majesty's Ship Sulphur, during the years 1836–1842, including details of the Naval Operations in China, from Dec. 1840, to Nov. 1841* by the ship's captain, Edward Belcher. The surgeon on board, Richard Brinsley Hinds, produced three companion volumes describing the mammalia, mollusca and flora they saw on the voyage.

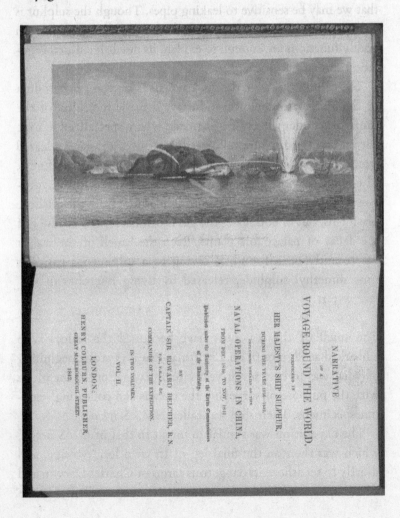

Belcher's *Sulphur*, a ten-gun bomb vessel, was the third of three Royal Navy ships of that name. The first of them already bore this curious name when the navy bought it from its American owners in 1778. I have been unable to discover a specific reason for its chemical christening. I suppose that it was simply regarded as a suitable sign of belligerence since the second HMS *Sulphur*, purchased in 1797, took part in the Battle of Copenhagen alongside sister ships named *Volcano*, *Explosion* and *Terror*. Like the second, the third *Sulphur* was equipped with mortars that were able to lob explosive shells or 'bombs' forward from the bow rather than simply firing cannon from the sides. This capability was to be put to use when the ship was sidetracked from its scientific mission into military conflict with China during the First Opium War.

Her route took the *Sulphur* via Tenerife and the Cape Verde Islands, around Cape Horn, and up the South American coast to Panama, from where she made three huge perambulations of the north and south Pacific, plumbing the depths and scanning the horizons for unknown islands, before passing westward through the Pacific islands and the Malacca and Madagascar straits and round the Cape of Good Hope to home. The main task was survey work, for which the ship was equipped with chronometers, both 'pocket' and 'heavy', and rockets that could be sent up to provide time signals. Comparing chronometer readings at two stations on land at the moment the rocket flare is seen allows the distance between them to be calculated. As the crew took readings on the island of Gorgona, off the Colombian coast, some faulty rockets went off at ground level. Fortunately, though, there was enough gunpowder to try again. A second signal was successfully improvised by hauling some bags of powder up a tall tree and igniting them there.

At Nootka Sound in British Columbia, 'Indians' gathered round the *Sulphur* in their canoes seeking to trade fish and furs.

Some form of entertainment seemed to be in order, and so Captain Belcher sportingly went ashore at dusk with 'a magic-lanthorn and supply of fireworks'. The lantern show inspired delight, but the fear occasioned by the fireworks was such that 'I had several women grasping me by each hand'.

The *Sulphur*'s voyage up to this point had amounted to a Cook's tour of some of the world's geological hotspots – the Canaries, Panama, the Sandwich Islands (Hawaii), Alaska. Belcher went yomping up Mexican volcanoes as if they were the Munros. Five thousand feet up on the rim of one of the three craters of the Viejo volcano, he plunged a thermometer into the soil and found the temperature went off the scale. 'It speedily warmed me to an unpleasant degree through thick boots.' Another time, at Tepitapa, where the Lake of Managua falls into the Nicaragua River, they dallied by a sulphur spring: 'My thermometer was not graduated above 120°, therefore I cannot state more than that eggs were boiled in it,' Belcher reported. 'Crystallization was abundant on the small stones between which it flowed, and some specimens I examined were a mixture of sulphur and calcareous matter. The taste was not unpleasant.' Belcher does not on this or any other occasion think it worth noting the coincidence of his vessel's name.

Meanwhile, Surgeon Hinds and his scientific assistants observed or collected whelks, clams and scallops, lemurs and jerboas, parrots and kingfishers, mimosas, euphorbias, cactuses and oaks. The discovery that sulphur plays a role in plant and animal life had been made a generation or two earlier from investigations of horseradish and ox bile. The men perhaps would have been aware of this, although they would not have known, for example, that their clams fed for bacteria around submarine sulphur vents. Nor were they fortunate enough, as the *Sulphur* passed through the Malacca Strait, to discover the Sumatran titan arum, or corpse flower, whose massive bloom bursts open once every several years to release a cadaverous stink based on a cocktail of dimethyl polysulphides.

But more sulphurous adventures lay in wait. At Singapore, Belcher picked up orders from the Admiralty to proceed instantly to Canton in order to take part in naval operations against the Chinese. The First Opium War had broken out in 1839, when Britain seized Hong Kong in an attempt to force China to open to trade. The *Sulphur*'s botanist on board made his excuses and left to return to Kew, 'conceiving himself out of his sphere in our prospective cruize'. On 7 January 1841, the *Sulphur* took up her position at the outer defences of the Canton River and began shelling the enemy, 'giving lower Chuenpee a dose of grape and canister'. Then they attacked the junks. One rocket set off the magazine of the ship closest to the Chinese flagship 'and she blew up in great style'. The British prosecuted their success by capturing an important fort, only to find that the enemy had removed its guns in the night. 'A "flare up" we might easily have had, as the lines were plentifully strewed with powder.'

On their return to Spithead, the surviving crew of the *Sulphur* were pleased to learn they had qualified for a bonus under a scheme brought in during their absence to compensate them for the longevity of their voyage. Edward Belcher was knighted. Richard Hinds opened his cases to find that many of his specimens had been 'reduced to powder' by insects and later learnt that 200 species of plants laboriously gathered from California and the Pacific islands were 'already described'.

What the circumnavigation of HMS *Sulphur* unwittingly demonstrated was the ubiquitous occurrence and daily utility of the element for which she was named. Her crew paid their respects at its frequent eruptions from the earth and put it to use for science, revelry and war. The ship returned to a country where the inventor Thomas Hancock had just obtained a patent for the use of sulphur in the vulcanization of rubber, and the brimstone terrors of the Book of Revelation had been suffi- ciently tamed that the name of Lucifer could be tolerated as a brand of matches.

### Pee is for Phosphorus

Long before phosphorus came to scientific light, there was Phosphorus, gentle usher of the dawn:

> Sweet Phosphor, bring the day!
> Light will repay
> The wrongs of night;
> Sweet Phosphor, bring the day!

So wrote Francis Quarles in his *Emblems Divine and Moral* of 1635, in a typical evocation of the morning star known in Greek as Phosphoros and Latinized as Phosphorus. The morning star we know now – and in truth it was known then, although poetic fancy favoured the idea of a self-generating light – is the planet Venus, which is always seen in the sky near the sun, and brightly reflects its light so to appear to us as the herald of the new day. The same planet does double duty as the evening star Hesperus, which catches the light of the sun just set, and to which the poets, natural late risers, refer rather more often than Phosphorus.

So useful were these luminous companions of the dawn and dusk that they continued to gain lyric employment long after science had shown the names to be falsely applied. Poetic Phos- phorus was not yet at its zenith when one Hennig Brand of Hamburg took it for the name of the new element he discov- ered, probably in 1669. Slowly, though, the poets began to assimilate the additional meaning. In the nineteenth century, for example, Tennyson's 'In Memoriam' still invokes Phosphor to indicate the time of day, as does Keats atop Ben Nevis. But in his poem 'Lamia', Keats is drawn to the idea that this natural glow might be captured by man – literally caged, in fact – as he describes a portal 'Where hung a silver lamp, whose phosphor glow / Reflected in the slabbed steps below, / Mild as a star in water'. This image corresponds in turn to accounts of 'perpetual lamps', which were presumably based on phosphorescent materials, supposedly used by early Christians such as Saint Augustine.

The idea of a substance glowing with light that seems not to be fire is a compelling one, and elemental phosphorus does indeed glow in the dark. The light comes from the combustion of short-lived oxides that are created at its surface when it is exposed to air, something only confirmed in 1974, 300 years after Brand first observed the eerie light. But not everything that we tend to describe as phosphorescent actually owes its glow to phosphorus. Marine phosphorescence – observed at night in warm waters when the sea blooms milky white like the negative of a photograph – occurs when enzymes trigger chemical reactions in bioluminescent bacteria and does not involve phosphorus itself. Similar chemistry is observed in other luminescent organisms from fireflies to honey fungus.

Phosphorus is involved in some equally peculiar goings-on, however. Herring, for example, are said to emit light as they rot away. Intrigued by this unlikely claim, I bought some herring, and set one out to decay in the garage where the ammonia smell would not be too oppressive. Two nights later, I felt my way towards where I had put the fish. At first, I saw nothing. But as my eyes grew accustomed to the blackness, I was amazed to detect the faintest of glows following the torpedo shape of the herring, the brightest part lying towards the head. In *The Rings of Saturn*, W. G. Sebald says 'this glowing of the lifeless herring' remains to be explained. But the chemistry is straightforward.

Along with the ammonia is generated a smaller quantity of its phosphorus analogue, phosphine, and a related compound, diphosphine, which is spontaneously combustible. The slow flame of this gas burning as it seeps forth from the fishy carcass is what produces the light. The same reaction has been proposed to explain tales of human spontaneous combustion. In *Bleak House*, Charles Dickens memorably has the rag-and-bone man Krook meet his end in this way. His lodger finds him a 'crumpled black thing' having suffered a death 'engendered in the corrupted humours of the vicious body itself'. Dickens reveals that he had read up on human spontaneous combustion when he cites a number of 'true' cases in his account of the inquest into Krook's death. When the episode first appeared in serial form, Dickens was criticized by the philosopher George Henry Lewes and others for giving credence to pseudoscientific ideas. But he stoutly defended his position in the preface to the published novel, adding a reference to a further case only recently reported. Puzzling stories of human spontaneous combustion still crop up from time to time, although first-hand witnesses tend to be thin on the ground. Phosphorus released by the body has not been eliminated as a possible source of ignition.

Hennig Brand was an alchemist who had married well, and with his wife's indulgence, he was able to maintain a laboratory on the Michaelisplatz in the shadow of the newly completed St Michaelis church in the new part of the thriving Hanseatic port of Hamburg. He was an upright, if somewhat pompous, citizen, earning the satirical nickname of Dr Teutonicus, although his real name now seems more appropriate than any alternative: Brand is the German for fire. He believed according to the alchemical orthodoxy that there might be a divine connection between the gold he sought and that abundant golden liquid, human urine. This led him to collect and evaporate a large quantity of urine and then to distil the residue. He noticed that the vapour which came off had a ghostly glow, and that the waxy white material which condensed from it had the same inner light. It also burst into flame as it escaped from the retort and came into contact with the air. He was astonished to find that the light was not dependent upon the heat of the experiment, but seemed to be an intrinsic property of the mysterious substance. Brand realized that he was now in possession of something quite remarkable, a miraculous light that sprang from the substance of our own bodies. Perhaps it was the philosopher's stone itself. At the very least, it had to be a sign. Diligent alchemist that he was, he spent the next few years in a futile attempt to convert his find into gold. Others sought to take advantage of Brand's success, but the philosopher Gottfried Leibniz, then in the employ of Duke Johann Friedrich of Hanover, befriended the alchemist and set up contracts for him whereby he was able in the end to obtain at least some gold for his efforts.

Brand's experiment – the earliest documented piece of science that led to the discovery of a new element, even if it was not appreciated as such at the time – looked like the sort of thing I ought to be able to repeat at home. I could make my own phosphorus from my own urine.

But first, if I was to have any chance of succeeding, I would need a more precise recipe. Where was this to be found? Brand did not publish his work, at first keeping it secret, only occasionally spilling crucial details in exchange for a few thalers. With these meagre clues, Brand's rivals were unable to repeat his achievement for some years. On the rare occasions when somebody succeeded, they too took steps to preserve the mystery: it naturally heightened people's interest to unveil a glowing specimen of the miraculous substance if one kept quiet about how it was made.

There are many paintings of the famous scientists associated with the discovery of the elements – foremost among them Jacques-Louis David's sumptuous portrait of the great chemical modernizer Antoine Lavoisier and his wife – but very few that show them at work, or that purport to show the moment of discovery. However, the discovery of phosphorus is an exception.

Of this there is a marvellous painting made by Joseph Wright. It bears the cunningly revealing title: *The Alchymist, In Search of the Philosopher's Stone, Discovers Phosphorus, and prays for the successful Conclusion of his operation, as was the custom of the Ancient Chymical Astrologers.*

I went to see the painting in the city art gallery of Derby, where Wright was born and worked for most of his life. There was much in it to wonder at. Why is the 'alchemist', Brand, wearing monkish robes and working in a vaulted Gothic room if this is 1669? The scene is more like a Frankenstein film set than a proper laboratory. Such anachronisms are probably deliberate, as we shall see. For the moment, though, I need to focus on the experiment in progress. Wright shows Brand kneeling, hands outspread in amazement, before a brightly glowing glass flask which sits on a tripod stool. Next to this is a plastered brick chimney flue, free-standing in the room, which is fed by an unseen fire. A pipe runs out from the top of the flue and down into the flask, and some luminous material is pouring along this pipe into the flask. It is clear that no heat is being applied to the flask, and that every effort has been made to exclude air from the apparatus, for the joint between the pipe and the flask is carefully sealed with clay. Both of these details emphasize that the generated light is to be regarded as a natural wonder and not as any sleight of hand on the part of the alchemist.

Wright's fanciful image could not be taken as reliable evidence, of course, but it seemed to offer encouragement. The set-up was simple and brutal, which would make it easier for me to replicate the experiment. And I now knew what I should expect to see if it worked. But the starting materials concealed within the furnace remained as mysterious as ever. How did one get from liquid urine to anything that could be put in a furnace?

Fortunately, as Brand and his competitors traipsed round the courts of Europe with specimens of the *noctiluca*, or 'night-light', stoppered up in their pockets, some of the leading scientists of the day were there to take notes and make their own investigations, from which more coherent recipes did begin to appear. One of the clearest is to be found in the papers of Robert Hooke, one of the original fellows of the Royal Society, published twenty-three years after his death, in 1726:

Take a Quantity of Urine (not less for one Experiment than 50 or 60 Pails full); let it lie steeping in one or more tubs, or an Hogshead of oaken Wood, till it putrify and breed Worms, as it will do in 14 or 15 days. Then, in a large kettle, set some of it to boil on a strong Fire, and, as it consumes and evaporates, pour in more, and so on, till, at last, the whole Quantity be reduced to a Paste, or rather a hard Coal, or Crust, which it will resemble; and this may be done in two or three Days, if the Fire be well tended, but else it may be doing a Fortnight or more. Then take the said Paste, or Coal; powder it, and add thereto some fair Water, about 15 Fingers high, or four Times as high as the Powder; and boil them together for ¼ of an Hour. Then strain the Liquor and all through a woollen Cloth; that which sticks behind, may be thrown away, but the Liquor that passes must be taken and boil'd till it come to a Salt, which it will be in a few Hours.

After that, it was simply a matter of adding some *Caput Mortuum* (or 'death's head', which, apparently, 'you have at any Apothecary's') to the salt and soaking the resulting mess in alcohol 'so as it will become a kind of Pap':

Then evaporate all in warm Sand, and there will remain a red, or reddish, Salt. Take this Salt, put it into a Retort, and, for the first Hour, begin with a small Fire; more the next, a greater the 3d, and more the 4th; and then continue it, as high as you can for 24 Hours. Sometimes by the Force of the Fire, 12 Hours proves enough; for when you see the Recipient white, and shining with Fire, and that there are no more Flashes, or, as it were, Blasts of Wind, coming from Time to Time from the Retort, then the Work is finished. And you may, with Feather, gather the Fire together, or scrape it off with a Knife, where it sticks.

The Fire is best preserved in a Vessel of Lead, closed up from the Air. But to be seen, 'tis also put into a Glass, in Water, where it will shine in the Dark . . .

This was beginning to sound epic. Fifty or sixty pails of urine was an awful lot for a start. How long would it take me to produce that much? In fact, I learnt, I should be able to take some short cuts and stand a chance of replicating the experiment on a smaller scale. One bucket of urine – about three days' worth – should contain around four grammes of phosphorus. This, if only I could separate it, would be more than enough to ignite 'the Fire'.

The first question was which urine to collect. The health guides always say it should be 'straw-coloured', as if everybody is intimately familiar with the colour of straw. Should I take this typical sauvignon blanc effluence? I decide it would be better to use the over-oaked chardonnay of the morning's first pee. This strikes me as more likely to be rich in dissolved substances. I collect four litres and allow it to evaporate in an open container which I put out in the garden. It reeks at first, but gradually the disgusting smell disperses, and the liquor turns a rich ale brown. I am relieved to see that it shows no sign of breeding worms, not only because I have no especial wish to pick them out of the putrifying concentrate, but also because it implies that my sample is uncontaminated with stray organic matter, and that I might therefore be able to omit some of the repetitious purifying stages found necessary in the seventeenth century. After several weeks baking in the sun, all the liquid has evaporated, and I am left with twenty-two grammes of an almost odourless crystalline residue the colour of sawdust. This, I hope, is the reddish salt observed by Hooke.

Now I am ready to begin the long roasting process. For this I need some more professional laboratory apparatus and advice. I enlist the help of Andrew Szydlo, one of my former chemistry teachers. Andrew is a man of many talents, and I remember him as always liable to whip out his gipsy violin in mid-lesson, or pass on some piece of lore about bee-keeping or car maintenance. More pertinently, he is an authority on alchemical history and the author of a treatise on Michael Sendivogius, the Polish alchemist who may have discovered oxygen in the early seventeenth century and contributed to its use in a pioneering manned submarine crossing of the Thames by the Dutchman Cornelis Drebbel in 1621. Andrew speaks with an English orotundity and the trace of a Polish accent and is wont to greet his former pupils as 'professor!' He is enthusiastic about the attempt to replicate this first isolation of a chemical element, and has set out various ingredients that may prove helpful in our quest, not least some best-quality gunpowder charcoal he has made from willow wood.

We grind up some of my urine residue in a pestle and tip it into a test tube for heating. This tube is connected to an apparatus that will allow us to collect any distillate and test any gases that come off. Volatile material, including any phosphorus, should condense in a second test tube, while the gases escape through a vent. We aim two bunsen burners at the base of the loaded test tube, turn them up to maximum heat and wait. At first, a little water vapour comes off, which is followed by thick yellow curls that look and smell a little like burning tobacco. 'Very curious,' says Andrew in his demoniacal way. 'It's the most bizarre experiment, I must say.' This vapour condenses as a tarry brown oil much like that which is produced when many forms of organic matter are burnt in this controlled fashion. At the vent, wisps of a white vapour appear. Could this be phosphorus pentoxide, the acidic combustion product of phosphorus? Litmus paper shows that unfortunately it is alkaline; another quick test with hydrochloric acid confirms that it is merely ammonia. We allow the solid remaining in the test tube to cool. It is now a dark slate grey. A flame test – a little of this solid dabbed on to a platinum wire and placed in a hot blue flame – reveals the characteristic yellow light of sodium and a fainter carmine red due to calcium. Andrew is now giving me a masterclass in analytical chemistry, which he intersperses with tirades against the parlous state of chemical education: how the school janitors are always seeking to clear away miscellaneous apparatus that they take to be scrap; how students are hardly allowed to do experiments for themselves these days; or, if they are, how the experiments must be designed so as to deliver a result before the end of the lesson – a constraint that slams the door on any slow-roasting chemistry such as this.

Is the sodium just ordinary salt – sodium chloride – or is it perhaps, as we hope, a phosphate or phosphite salt, which would mean we're getting closer to our goal? Dissolving a little of the grey residue in water and adding a drop of silver nitrate rapidly produces a dirty precipitate. This separates into a milky white sludge – the standard proof of chloride – and a mysterious brown residue, which does not dissolve either in acid or alkali, suggesting that it is rich in inorganic substances. This is probably where the phosphorus still lurks. We decide to reheat the residue mixed with the special charcoal Andrew has brought in order to reduce the phosphate or phosphite to elemental phosphorus. We grind the two materials together – the grey burnt urine residue and the black willow charcoal – and subject the mixture to the bunsen burners. 'We're roasting the pants off this now,' he says gleefully.

I am surprised when the residue, which has already spent an hour or so at the highest temperatures we can produce in a school laboratory, begins to react anew. Andrew explains that by grinding it with the charcoal we have greatly increased the surface of contact between the two materials, so boosting the chances of a reaction. More ammonia comes off, followed by a gas that burns with a low blue flame when a taper is put to it. It is now dusk, and we turn out the laboratory lights in order to study the flame more closely. Could this be our phosphorus? It cannot be, because it would produce a thick white smoke of phosphorus pentoxide. It is presumably carbon monoxide burning to the invisible carbon dioxide. As the flame dwindles away in the darkened laboratory, it seems to betray a faint white edge in its dying moments. 'We may just be beginning to get something,' Andrew tells me. Limited now by the temperature – five or six hundred degrees Celsius – achievable with bunsen burners, we are brought up against the knowledge that Brand and his imitators used far hotter furnaces and ran the experiment for hours or days. We resolve to meet again, armed with quartz test tubes and an oxyacetylene torch that will enable us to turn up the heat.

This time, it is immediately clear we are reaching a much higher temperature. The sequence of observations that we previously noted over an hour or more is repeated within minutes. Very soon, the roasted residue in the quartz tube begins to glow with a dazzling white light. Excitedly, we assume this may be our phosphorus, but the glow stays resolutely at the tip of the turquoise oxyacetylene flame where the heat is greatest. If it was really phosphorus, it would flow out of the tube in a vapour which would condense in the cooler second tube, as in Wright's picture. It seems it is merely an incandescence produced by the extreme heat as it vaporizes the very substance of the quartz tube. We are forced to concede that, whatever his delusions, Brand was clearly a formidable experimental scientist.

Joseph Wright of Derby painted *The Alchymist* in 1771. It was one of a number of scientific demonstrations he committed to canvas: his most famous work is probably *An Experiment upon a Bird in the Air Pump* done a few years earlier, in which a well-to-do household gathers in various states of wonder, horror and pity around a glass bulb from which the natural philosopher who stares unflinchingly out at us from the centre of the composition has evacuated all the air, extinguishing the life, or at least the temporary consciousness, of the bird inside.

Wright was well connected with the Lunar Society in nearby Birmingham, whose members, including James Watt, the inventor of the steam engine, the physiologist and poet Erasmus Darwin and the chemist Joseph Priestley, met mostly on the full moon so that they could see their way home after evenings of 'dinner and a little philosophical laughing' that on occasion also included an experimental demonstration. Inspired by Robert Boyle's work on vacuums in the 1650s, the painting seems also to anticipate Priestley's experiments on the life-affecting properties of the new gases oxygen and carbon dioxide that lay a few years in the future – and the full moon shines in through the window. Other members of the society, such as the industrialists Josiah Wedgwood and Richard Arkwright, bought his work. With these paintings, Wright made his name as a recorder of the scientific Enlightenment.

Like *The Air Pump*, *The Alchymist* is reimagined history. It purports to show the first making of phosphorus, which also happened more than a century before. Interpreted as an allegory, the painting seems to represent modern science shining its light into the alchemical darkness, a message that would naturally go down well with Wright's patrons. However, the work appealed neither to them nor to contemporary viewers; it was unsold at Wright's death in 1797. An astute analysis by the art and science historian Janet Vertesi attempts to explain its 'curious failure', and to account for the weird get-up of the protagonist. The painting balances three sources of light – once again the full moon outside, the radiant phosphorus pouring into the flask, and, on a bench in the background, the dimmer light of an oil lamp by which two laboratory assistants are going about their own business apparently oblivious of the miraculous scene unfolding before them. This trinity of lights may have religious significance, but it also symbolizes a contest between nature (the moon), the Enlightenment (the oil lamp) and some mysterious, more powerful, third force. The rational students of nature (they are in modern dress and use modern apparatus in contrast to their druidical master) toil by lamplight, but they are outshone, literally, by the light of the ignorant alchemist's *accidental* discovery. Recall Wright's carefully phrased title: 'The Alchymist, In Search of the Philosopher's Stone, Discovers Phosphorus . . .' In other words, the alchemist, while doing whatever it is that alchemists are supposed to do, inadvertently makes a genuine contribution to science, a contribution moreover that the rationalists have failed to make for themselves.

What kind of message did this send to the Enlightenment pro-gressives of the Lunar Society in the fast-industrializing English Midlands?

Science had the last laugh, however. Brand and the few rivals who eventually managed to repeat his experiment toured the courts of Europe with their precious luminous cargo. In Eng-land, Charles II attended a demonstration, as did Samuel Pepys and his fellow members of the Royal Society. John Evelyn wrote of how, while dining with Pepys in 1685, they witnessed 'a very noble experiment' in which two liquids were mixed to produce 'fixed divers suns and stars of real fire, perfectly globu-lar, on the sides of the glass, and which there stuck like so many constellations, burning most vehement'. But for a long time phosphorus remained little more than a high-end party trick. Obtaining it was arduous and obscure, and its elemental status was far from agreed, with chemical dictionaries sometimes list-ing it as no more than a 'species of sulphur'.

Exactly 100 years after Hennig Brand isolated phosphorus from urine, the Swedes Carl Scheele and Johan Gahn showed that it was a major constituent of bone. This richer source of the element made it possible at last to consider how it might be put to practical use. For more compelling than a mysterious light in nature is, as Keats observed, a light that may be captured by man. By the time that Keats was writing 'Lamia' in 1819, phos-phorus lamps such as he describes were the latest thing, inventors having found a way to prevent the outright combustion of the phosphorus by diluting it in a suitable inert medium and regu-lating the admission of air. In this way, they were able to obtain a lamp that could give a steady glow on demand over a period of weeks. The discovery and application of phosphorus was well timed for the element to become a symbol of the taming of nature, of progress, and, literally, of enlightenment.

*

The British returned Hamburg's chemical gift to the world with vengeful interest during the last week of July 1943. In nightly raids, hundreds of aircraft dropped 1,900 tonnes of white phosphorus incendiary bombs on the city, the culmination of a strategy of 'morale bombing' authorized in 1941 by the prime minister, Winston Churchill, and Arthur Harris, the chief of the Royal Air Force Bomber Command, who sought to direct the aerial assault on locations most likely to weaken the spirit of the enemy. Increasingly, the manner of the bombing became a factor too, so that by the summer of 1943 the Allies' objective was to destroy cities not only of historic and industrial

importance, but also those densely populated with key workers, and to use means specifically designed to terrify the Germans into submission. This led to an unprecedented emphasis on incendiary bombs, and especially on phosphorus.

On 27 July, the third night of the onslaught, incendiary bombing combined with the hot, still weather to produce a firestorm, a phenomenon where the intensity of the conflagration sucks in air from all directions, so feeding the flames and creating a ferociously hot vortex of fire. In the words of a recent analysis by a German historian:

The combination of the climate, the incendiary ratio, the collapsed defenses, and the structure of the city blocks created what Harris's codename 'Gomorrah' predicted: Like Abraham in Genesis 19:28, Harris looked toward the sinful city 'and beheld, and lo, the smoke of the country went up as the smoke of a furnace.' It melted between forty thousand and fifty thousand people.

Many others were asphyxiated as the upward rush of flame simply sucked the air out of their underground shelters. Although the old town survived, the fires devastated much of the rest of Hamburg-Mitte, the central district where Brand had first isolated phosphorus nearly 300 years before. More than a quarter of a million dwellings were destroyed, along with factories, shipping and the all-important U-boat docks. Fifty-eight churches were reduced to rubble, but although its neighbourhood was badly hit, St Michaelis survived for another year until it was badly damaged in an American bombing raid. That autumn the trees of Hamburg flowered again as if it were spring.

'Dropping phosphorus bombs on the homes of innocent civilians is never likely to happen again,' writes John Emsley, while explaining that the element is nevertheless bound to remain a part of modern armouries because of its sheer versatility, used to illuminate targets, to create smokescreens or to ignite and clear vegetation. Yet as I write, in January 2009, Israel has admitted using white phosphorus during its offensive in Gaza. Israeli fire first hit a United Nations school, and a week later officials at the United Nations Relief and Works Agency for Palestinian Refugees in the Near East claimed that its Gaza City compound was set ablaze by phosphorus shells. In this conflict, as in others since the First World War, phosphorus is regarded as a legitimate agent of warfare, but its use is confined by convention to the open battlefield, and it is not allowed to be used against civilian populations. In Gaza, it happened that the 'battlefield' was densely populated: the smokescreen that phosphorus produces remains moral as well as literal.

## 'As under a green sea'

The red poppy, which we wear to commemorate the loss of life in the First World War, offers us consolation because it is a symbol of survival: a flower that grew from the soil of battlefields, which had been fertilized with the blood of the slain. But one of the weapons of that war destroys even this sentimental construction. The poison gas that both sides employed offensively for the first time in 1915 had horrific power to choke the lungs, and to bleach the grass and flowers white. The gas was chlorine.

By the outbreak of the war, it had been anticipated for some fifty years that new chemical weapons, built on the scientific advances of the nineteenth century, would be developed for use in war. However, so strong was this possibility, and so strong too the sense that this was something uniquely abhorrent, that a pre-emptive ban had long been in place to regulate the use of such lethal agents on the battlefield.

Tear gas remained legal because it did not kill. The challenge for military engineers was to find a means of delivering it to enemy lines on a large scale and ensuring it would disperse in such a way as to cause maximum disruption while presenting minimal danger to one's own forces. The German chemist given this job was Fritz Haber – the same Haber who would later endeavour to extract gold from seawater for his country and who was already celebrated as one of the innovators of a process for converting nitrogen from the air into ammonia. When, later, he was awarded the Nobel Prize for this work, the choice was highly controversial because he had by then been listed by the Allied Powers as a war criminal.

Haber's brainwave was to keep it simple. Chlorine was a step backwards from tear gas in chemical sophistication, but a considerable leap forward in practicality. Rather than try to encapsulate it in shells that could be fired behind enemy lines, Haber proposed simply to release the gas from ground-based cylinders and allow the wind to do the rest. The chlorine, twice as heavy as air, would roll along the ground in a choking blanket before which the enemy would have no option but to retreat. At Ypres in northern Belgium, Haber personally oversaw the installation of more than 5,000 cylinders along a seven-kilometre stretch of the Western front. Chlorine became the first weapon of gas warfare on the afternoon of 22 April 1915, when a light north-easterly wind blew favourably for the German army. The surprise attack seemed to engulf the Allied soldiers, mainly French and Algerians. Swallowed up by the corrosive cloud, they could no longer tell whether to retreat from the gas or push through it in the hope of finding clear air beyond. By the end of the day, hundreds of men lay dead, and thousands were incapacitated, many remaining permanently disabled.

Did chlorine break The Hague conventions banning 'asphyxiating and deleterious' substances? Haber's argument that chlorine was non-lethal, like tear gas, and so a legitimate weapon of war, seems disingenuous in the light of his infamous later boast that he had devised 'a higher form of killing'. The death toll on that April afternoon in Ypres pronounced its own verdict.

Certainly the attack was felt to be more than enough to sanction response in kind by the Allies. Both sides employed gas periodically throughout the rest of the war, although it was never so devastating as when used by the Germans at Ypres and a few weeks later on the Eastern front west of Warsaw. Both sides also displayed an alarming readiness to deploy ever more unpleasant gases, escalating the chemical war with agents such as phosgene (carbonyl chloride), which smells faintly of fresh hay, mustard gas and other chlorinated compounds of sulphur and arsenic. But it is chlorine that still seems the most brutal weapon because of its elemental simplicity. The gas rips through the blood vessels that line the lungs and the victim eventually drowns in fluid produced as the body attempts to repair the damage.

Haber's patriotic efforts cast a long shadow, not least over his own family. His wife, Clara, committed suicide on the night of 1 May 1915 using her husband's service revolver. Biographers argue over the extent to which the death was her protest against Haber's chemical war, but it is worth noting that she was a qualified chemist herself, having trained in the subject in order to catch Fritz's attention, and had observed the effects of chlorine in Haber's animal experiments and field trials. Haber was apparently unfazed; he departed the following morning to supervise the installation of gas cylinders on the Eastern front.

Haber's son by his second marriage, Lutz (a contraction of Ludwig-Fritz), was haunted by his father's history and tried to lay the ghost in a book called *The Poisonous Cloud*, which remains one of the standard works on chemical warfare. Haber was forced to leave his beloved Germany with his family when his Berlin research institute was shut down by the Nazis in 1933.

(Although his chemical talents would undoubtedly have been put to use, and indeed he offered his services, Haber's partly Jewish ancestry made him unacceptable.) He considered settling in Palestine, then looked to find a home in Cambridge; in the end there was time for neither, and he died just a few months after his journey into exile.

Lutz Haber and his sister Eva Charlotte stayed in England. A number of years ago, I visited them in an incongruously genteel cottage in Bath, the city to which they had retired. Lutz was then nearly eighty and a little frail, but Eva Charlotte was the kind of woman who seems to save her sharp focus for old age. They remembered their father dimly – the odd game of bowls, helping him up the stairs, that sort of thing. Eva recalled how Einstein, a family friend, explained relativity to her by analogy with moving trains, and told me how she and Lutz one day climbed up a ladder and into Haber's institute, tripping over an assistant's apparatus and breaking it, sending their father into a rage. Why did Lutz write his magnum opus? 'I felt I should contribute my bit,' he confided. In his 'personal introduction' to the book, he elaborates with a critical sketch of his father as 'the embodiment of the romantic, quasi-heroic aspect of German chemistry in which national pride commingled with the advancement of pure science and the utilitarian progress of technology'. He judges his father's patriotism 'unusual even in an age when jingoism, into which it so frequently spills over, was condoned'. As for chlorine, Lutz told me, it was simply 'the most readily available substance. The chemical industry was very capable of producing chlorine quickly and in quantity.'

Wilfred Owen uses a chlorine gas attack as the tableau against which to expose 'The old Lie' of patriotism in the most famous poem of the First World War:

Gas! GAS! Quick, boys! – An ecstasy of fumbling,
Fitting the clumsy helmets just in time;
But someone still was yelling out and stumbling
And flound'ring like a man in fire or lime . . .
Dim, through the misty panes and thick green light,
As under a green sea, I saw him drowning.

In all my dreams, before my helpless sight,
He plunges at me, guttering, choking, drowning.

If in some smothering dreams you too could pace
Behind the wagon that we flung him in,
And watch the white eyes writhing in his face,
His hanging face, like a devil's sick of sin;
If you could hear, at every jolt, the blood
Come gargling from the froth-corrupted lungs,
Obscene as cancer, bitter as the cud
Of vile, incurable sores on innocent tongues, –
My friend, you would not tell with such high zest
To children ardent for some desperate glory,
The old Lie: *Dulce et decorum est*
*Pro patria mori.*

Owen limns the effects of the gas with a pathologist's accuracy. John Singer Sargent's famous painting *Gassed*, completed
after the war in 1919, confronts us with none of this frenzied
horror. His vast canvas shows a walking column of eleven
men, all but the man who is guiding them blindfolded and
holding on to the shoulder or knapsack straps of the man in
front. A similar column is led off by white-coated men in the
distance. Around the walking wounded other injured men lie
on the ground, one drinking from a water bottle, another
clutching a hand to his bandaged eyes. The bleak, flat landscape

is punctuated only by the pitched canvas of the nursing stations. Over it all, a low rancid sun forces its light through a
greenish sky.

There is clearly something wrong with this picture. It may
not be a picnic, but the scene is curiously static, almost repose-
ful. The soldiers are not suffering. There are no visible injuries,
no scarred or burnt skin, no blood; uniforms are neatly in place.
There is no sign of the choking that Owen describes. The picture was painted following a visit that the artist made to France
in the summer of 1918. The gas at this late stage of the war is
more likely to have been mustard gas, although the sickly
greenish mist hints at chlorine. The artist has clearly responded
to the official brief he was given to emphasize soldierly com-
radeship, but he cannot have painted what he saw if he saw the
aftermath of a gas attack. His giant mise en scène parades its
identikit blond Aryan heroes – the sons of the society women
whose portraits he had grown rich by, perhaps – like an heroic
film in Cinemascope.

In the light, airy calm of the reading room at the top of the
Imperial War Museum, I read letters home from Ypres and find
the same scene painted in very different colours. Sergeant Elmer
Cotton of the Fifth Battalion Northumberland Fusiliers
described how

The flat country all around was covered to a height of from 5 to 7 feet with a greenish white vaporous cloud of Chlorine gas . . . further on we passed a dressing station – propped up against a wall were a dozen men – all gassed – their colour was black, green and blue, tongues hanging out and eyes staring. One or two were dead and the others beyond human aid, some were coughing up green froth from their lungs.

I read other letters that speak of the confusion produced by the novel weapon ('a great stream of Sulpher', according to Infantryman James Randall; carbon monoxide according to an erroneous first report of the Ypres attack in *The Times*); of the Allies' unpreparedness (the British have 'bicarb of soda or something as an antidote', wrote Lieutenant-Colonel Vivian Fergusson); and of the effects from a Canadian nurse named Alison Mullineaux, who tended two men 'both lungs burnt out in each of them', the doctor himself having to leave the ward in order to vomit from the gas he breathed off the patients.

Chlorine's pungent nature was noted from the outset. The Swede Carl Scheele was the first to isolate the gas in 1774, noting its green colour, choking power and bleaching effect on litmus paper and plants. He made the discovery in pursuit of one of the great chemical projects of the day: to confirm whether or not all acids contained oxygen. Well-known acids such as sulphuric and nitric were known to contain oxygen. Hydrochloric acid, then known as muriatic acid (after *muria*, the Latin for brine), was a mystery. Antoine Lavoisier even called it oxymuriatic acid, believing its acidity had to be linked to oxygen. Scheele succeeded in obtaining chlorine during the course of his own experiments with this acid. However, this did not prove the absence of oxygen. It was not until 1810 that this was done by Humphry Davy, who confirmed that the gas Scheele had isolated was indeed an element, by combining muriatic acid with his own newly discovered metal, potassium, and obtaining from the reaction only potassium chloride and hydrogen gas – no oxygen.

Chlorine's propensity to combine with other elements to form hazardous new compounds, such as mustard gas, was also noted early on. One of these substances was the highly explosive liquid nitrogen trichloride. When Pierre-Louis Dulong first made this compound in 1811, it cost him an eye and three fingers. André-Marie Ampère warned Davy about the dangers, but Davy repeated the experiment anyway, and received cuts to his eye from flying glass.

The critic John Ruskin was sufficiently struck by the contrast between placid nitrogen gas and its explosive chloride to cite them figuratively in his 1860 essay *Unto This Last*, making an argument in favour of 'accidentals' and against man's complete control of his materials:

We made learned experiments upon pure nitrogen, and have convinced ourselves that it is a very manageable gas: but behold! the thing which we have practically to deal with is its chloride; and this, the moment we touch it on our established principles, sends us and our apparatus through the ceiling.

The dangerous compounds of chlorine that are more familiar today are those that have become notorious pollutants of the environment. Some of these have their origin in the research done by Haber and his colleagues. The ceaseless search for 'higher forms of killing' had consequences for species other than man. DDT was one by-product of this research, its efficacy as a pesticide identified during the course of laboratory tests on insects of potential warfare agents. DDT is a chlorinated hydrocarbon, a class of compounds in which chlorine atoms are substituted for hydrogen atoms on a carbon backbone. The herbicide known as Agent Orange used as a defoliant during the

Vietnam War is another. The group of refrigerant gases known as CFCs – chlorofluorocarbons – are others.

Chlorine is a Janus-faced element. It is abundant in nature – not least in the salt of the oceans – and is essential for life, playing an important role in regulating body functions. Like sulphur and phosphorus, it is usually safe enough in natural combinations. But when it slips its leash it can do great harm. This is what happened in the case of CFCs, the famously inert compounds originally adopted as a safe alternative to existing aerosol propellants and refrigerant gases. High in the stratosphere, sunlight strips out their chlorine atoms, setting up a chemical cycle that allows them to rampage through the ozone layer, breaking it up molecule by molecule.

Released in controlled doses, however, chlorine has the power for good. Our awareness of the stinging smell of chlorine gas comes not from the battlefield, but from public swimming pools where it is used as a disinfectant, from the bleach under the kitchen sink, and from the medicine cabinet and preparations such as TCP and the chloroquine tablets we take on exotic holidays. It is said that chlorinated drinking water brought to troops in the First World War saved more lives than were lost to the gas as a weapon.

As early as 1785, Claude-Louis Berthollet, a follower of Lavoisier and an inspector of dyeworks, published an account of his experiments with the new element. Adding to Scheele's observation that the gas had a bleaching effect, he showed that it was possible to make a safe and practical bleaching agent by mixing potash – potassium carbonate originally obtained from wood ash – with chlorine water. Berthollet's discovery was well timed. Bleaching cloth had traditionally been a laborious business, involving repeated washing and then prolonged exposure to sunlight, a procedure that took some months even in favourable weather. The common sight of fields laid out with sheets of linen inspired some memorable images, especially in Dutch art, such as the painting attributed to Jacob van Ruisdael of the bleaching fields outside Haarlem. (The cultural memory of white rectangles papering the landscape was later perhaps an inspiration to the abstract painter Piet Mondrian.) The Industrial Revolution led to an increase in textile production and a demand for a faster bleaching technique. Berthollet advertised his discovery to British scientists, and in 1786 the leading industrialists of the day, James Watt and Matthew Boulton, travelled to Paris to see Berthollet demonstrate his instant bleaching process. Watt discussed his steam engine with admiring French academicians and brought home information about Berthollet's process, which he then applied in his father-in-law's textile factory.

Like Odysseus' sulphur, chlorine was also soon recommended in the fight against infection and disease. The gas was awkward to administer, however, and always unpleasant, and for a long time was not a popular treatment. It was the devastating epidemic of influenza immediately after the First World War that helped to make chlorine acceptable – a double irony since the gas so recently used for killing men was not actually effective against the flu virus. When Calvin Coolidge, the most inert of American presidents, underwent chlorine-inhalation therapy for a cold over three days in 1924, the *Washington Post* headlined: 'CHLORINE GAS, WAR ANNIHILATOR, AIDS PRESIDENT'S COLD. Coolidge Much Relieved After 50 Minutes in Airtight Chamber.' Over-the-counter chlorine remedies began to proliferate. One ointment, Chlorine Respirine, applied to the nostrils 'liberated pure chlorine gas'. The product advertising burbled: 'Its discovery is, in fact, one of the greatest triumphs of science.' In 1925, the president's health presumably restored, the *Post* joyously drew the bigger picture: 'Chlorine to save more lives a year than its war toll.'

I am indebted for some of these insights into chlorine's properties to an unusual book that is in effect a biography of the
element, but is more notable as the permanent record of an
intriguing pedagogical experiment. Two lecturers in the history
of science at University College London asked their undergrad-
uate students each to explore a different aspect of chlorine's life
'in science, medicine, technology and war'. The project was
completed over several years, with students inheriting work
from their predecessors, adding to it and improving it little by
little until a unique chemical commonplace book had been built
up. The copy I borrowed from the library had never been
opened. Was it just fancy that I sensed a whiff of chlorine rising
from its fresh bleached pages?*

## 'Humanitarian nonsense'

In Stanley Kubrick's blackest of black comedies, *Dr. Strangelove*,
the paranoid American general Jack D. Ripper, holed up on
Burpelson Air Force Base, under siege by his own men, finally
reveals to the hapless RAF officer Lionel Mandrake why he has
launched the nuclear attack on the Soviet Union that will lead
by the end of the film to the destruction of human civilization.
'Do you realize,' he says, chewing mightily on his cigar, 'that
fluoridation is the most monstrously conceived and dangerous
communist plot we have ever had to face?' Ripper, it should be

---

*Almost certainly. I learn from the printers of this book that chlorine will
probably not have been used for bleaching the paper, and that even if it was,
there would be no residual odour on the pages. Curiously, though, you may
still be able to catch a whiff of the battlefield here: recent Finnish research has
suggested that the distinctive 'new book' smell may arise from hexanal, an
organic by-product of the paper-making process, which, like phosgene,
smells of new-mown grass.

said, is driven by a pathological fear of contamination of his 'precious bodily fluids', something that first came to him 'during the physical act of love'. As machine-gun fire rakes through his office, he explains that fluoridation began in 1946: 'How does that coincide with your post-war Commie conspiracy? Mandrake, do you realize that in addition to fluoridating water, why, there are studies under way to fluoridate salt, flour, fruit juices, soup, sugar, milk, ice cream? Ice cream, Mandrake, children's ice cream.'

The halogens, of which the element fluorine is the first and most reactive example, have quietly insinuated themselves into our lives. Like a night nurse, they go about their business dosing us up without our agreement, muttering as they go, 'it's for your own good'. Water is chlorinated and fluoridated, bromides are prescribed, table salt is iodized. We are never consulted, but we know the words. These simple medicaments have a primal quality that encourages us to reach for them as readily as we once reached for hyssop or rue. The bromide, or Bromo-Seltzer, features in hard-drinking American literature almost as often as the bourbons and martinis whose effects it is there to assuage. In Tennessee Williams's *A Streetcar Named Desire*, the alcoholic Blanche DuBois clutches her head and announces to nobody in particular: 'Sometime today I've got to get a bromo.' In Ernest Hemingway's *The Snows of Kilimanjaro*, a man dies at length on the mountainside because he has failed to put iodine on his injured leg. The cause of death, it is made clear, is not the original accident but the man's failure to apply treatment; it seems he subconsciously chooses death because it offers him an escape from that worst of Hemingway fates, the forming of a mature human relationship. Iodine was a miraculous disinfectant but it delivered a salutary sting. 'No humanitarian nonsense about iodine,' as the cynical adventurer Mark Staithes winces approvingly in Aldous Huxley's *Eyeless in Gaza* while being treated for a similarly testing injury. Leonard Cohen's 1977 song 'Iodine' gains its sense from this womanish contrariness of the element in medicine – stinging one minute, soothing the next.

General Ripper was right about one thing. Fluoridation did begin in America just as the Second World War was ending. In December 1945, Grand Rapids, Michigan, became the first city to be supplied with fluoridated water. A nearby city was designated as a control in what was to be a ten-year trial of its long-term effects on dental health, but fluoridation was pre-emptively declared a success and rapidly extended to other metropolitan water supplies including the control city, thereby ruining the experiment. Well over half of Americans drink fluoridated water today – as near to free universal health care as that country comes, perhaps. The programme has been resisted by the libertarian John Birch Society and many other lobby groups. Accusations of conspiracy have been traded ever since: that fluoridation was a scheme solely dreamt up so that the aluminium industry could dispose of the large quantities of fluorine compounds used in the metal's manufacture; that it was funded by the sugar industry to get themselves off the hook for rotting people's teeth; and, because fluoridation in McCarthy-era America was backed by the government, that it was the *anti*-fluoridationists, ironically, who were the left-wing stooges. The objection on principle has been mostly not about the efficacy of fluoride in preventing dental disease, but directed towards the cavalier attitude of officialdom in compulsorily imposing blanket 'treatment' without the normal medical precautions of prior diagnosis, prescription and determination of dosage. Some European countries have discontinued water fluoridation and have introduced optionally purchased fluoridated salt and toothpaste in its place. Meanwhile, the people of the United States still unexpectedly comprise one of the most comprehensively fluoridated populations in the world, and controversy continues to flourish, with one typical website terming fluoridation 'medically evil, as well as socialistic'.

There never was a campaign against bromides, salts once so widely employed as all-purpose sedatives that the very word still retains humorous connotations of sexual underperformance. Although they had enjoyed great favour, they were withdrawn from the American market without fuss in 1975. By this time, so many dangerous side-effects had come to light that they earnt their own diagnostic description: bromism.

Bromides had begun to gain a reputation as a remedy more than a century before. In 1857, Sir Charles Locock, the physician-accoucheur who had attended Queen Victoria at the birth of her nine children, having heard that epileptic patients treated with bromide also experienced reduced libido, decided to try it on women suffering from 'hysterical' disorders. The amusingly named Locock shared the expert opinion of the time that epilepsy was linked to masturbation, nymphomania and other manifestations of 'excessive sexual excitation', and reasoned that, since his women appeared to be at their most disturbed during menstruation, the bromide treatment might also be an effective way to suppress the lustful desires that were supposedly troubling them. Proven to be effective as both anticonvulsant and anaphrodisiac, bromide seemed to confirm a link between epilepsy and onanism, and it began to be prescribed wherever a general dulling action was required. When the American humorist Gelett Burgess divided the world into two types, the sulphites and the bromides, in his 1907 book *Are You a Bromide?*, the term was widely understood to refer to a bore – sulphites, presumably, were people who by contrast brought a certain pungency to the table.

The same salt, potassium or sodium bromide, was also the active ingredient in the 'bromos' called for by Blanche DuBois, W. C.

Fields and other high rollers. The generic term evolved from Bromo-Seltzer, a commercial antacid sold in the form of a fizzy powder developed by Captain Isaac Emerson of Baltimore, Maryland. The splendid Florentine Bromo-Seltzer Tower still stands in the city, the twelve positions on its clockface spelling out the name of the drug. The brand persists although the product no longer contains any bromine, while the tower has been converted into writers' studios where a new generation can nurse their hangovers.

Iodine, though the elemental equal of fluorine, chlorine and bromine, appears to us not only as less dangerous than its

halogen fellows but even as something of a beneficence. Iodized salt is just as widespread in America as fluoridated water, yet its introduction from the 1920s never aroused libertarian passions. Its familiar medicinal form is tincture of iodine, simply the element in alcoholic solution. Brown liquid in a brown bottle, it seems pure unction, its heady aroma and staining colour like a kind of vanilla essence for external use only.

Iodine is one of the great accidental discoveries of science. In 1805, Bernard Courtois took over the running of his family's loss-making saltpetre factory in Paris while his father was in debtors' prison. Although the Napoleonic Wars had begun, Paris was at peace after the years of revolution, and there was little local demand for his explosives. Nevertheless his raw material, in particular the guano from which saltpetre is most conveniently made, was becoming increasingly hard to obtain. Courtois struggled to keep the business going, preparing saltpetre (nitrate of potassium or sodium) instead from wood ash. When even wood ash ran short, he turned to seaweed, traditionally harvested from the Brittany and Normandy coast for its soda, which was used in the making of glass. One day in 1811, he noticed some corrosion of the copper vessels in which he mixed seaweed ash with other ingredients to make the saltpetre. By experiment, he found that the pitting arose during the furious reaction that occurred when sulphuric acid was added to the alkaline soda. This reaction, he couldn't help noticing, also released puffs of an entrancing violet vapour. Investigating further, Courtois found that the vapour did not condense to a liquid but formed unfamiliar black, metallic-looking crystals. Courtois suspected he might have discovered a new element, but lacked the equipment to make tests and could not afford to take the time from his business. Instead, he asked two friends to finish the work. One of them, the gas chemist and balloonist Joseph-Louis Gay-Lussac, proposed the name iodine in analogy to chlorine.

By a strange chance, Humphry Davy was also present at the christening if not the birth. Since 1792, it had been difficult for British travellers to enter France, but Davy, who had been awarded the Napoleon Prize, was personally granted a passport by the emperor in order that he might collect his award. In October 1813, the newly married Davys, with a nervous young Michael Faraday acting as their footman, embarked at Plymouth on a ship used for the exchange of prisoners of war and set sail for Brittany. After a rainy voyage, they landed in enemy territory and were searched, even including their shoes. As they progressed towards Paris, they found the kitchens filthy but the food surprisingly agreeable. Davy had high hopes 'through the instrumentality of men of science, to soften the asperity of national war', but seemed unwilling to make the first move: at the Louvre he averted his eyes from the paintings lest he felt obliged to pay a compliment to his hosts. Jane Davy, meanwhile, shocked passers-by in the Tuileries Gardens with her unfashionably tiny hat.

Davy met with Ampère, his correspondent who had warned him of the dangers of nitrogen trichloride, and who had obtained some of Courtois's new substance. Using his travelling set of chemical apparatus, Davy subjected it to analysis and concluded with Gay-Lussac that it was indeed a new element, and related to chlorine. Davy annoyed Gay-Lussac by firing off a paper to the Royal Society to this effect, while Davy felt the Frenchman had merely asked him in the first place in order to pick his brains. However, it was all smiles when, towards the end of his two months in Paris, Davy was honoured to be made a corresponding member of the French Academy of Sciences. The Davys did not meet Napoleon himself, but they did visit the Empress Josephine at Malmaison before pressing on to Italy, Switzerland, Austria and Germany, returning home in April 1815, a few weeks before the Battle of Waterloo. Somewhere en route, Davy must have revised his opinion on the 'asperity of national war', for soon afterwards he wrote to the Prime Minister, Lord Liverpool, urging severe treatment for the French under the terms of the peace treaty.

After 1815, as the demand for saltpetre fell still further, Courtois sought to profit from his discovery of iodine, manufacturing the element and various compounds, using chlorine gas to displace the iodine in the liquor obtained from kelp ash. But he was again unlucky, soon overtaken by more efficient processes. Fame ultimately eluded him, and he died penniless in 1838.

Following Courtois's discovery, iodine was soon identified in seawater and in various mineral sources and was recognized to be effective in treating goitre. This revelation explained the traditional remedy of using burnt sponges or kelp to treat the swelling. The kelp ash industry that had been established along the weed-strewn rocky coasts not only of northern France but also of western Scotland had gone into decline when vast deposits of soda and potash were discovered in Spain and South America, but now it enjoyed a brief revival producing iodine for medicine. This business provided a meagre subsistence for crofters who kept kelp fires burning summer long to produce the iodine-laden ash. Entrepreneurs sought to put this activity on an industrial footing, with Glasgow becoming the centre. In 1864, the first factory on Clydebank was one erected in order to process thousands of tonnes of kelp brought up the river each year from the Scottish islands. But in an echo of what had already befallen the saltpetre industry, this labour- and energy-intensive process became uneconomic overnight when iodide deposits were found in Chile.

Although my nearest coastline is the flat sand and mud of East Anglia, where the seaweeds are not so lush as on rockier shores, I decide I should have a go at making my own iodine. I read careful instructions that I should select only this kelp or that laminaria, but, slipping among the tidal pools on a freezing December day, it is hard enough to distinguish one species from another at all. With numb hands, I randomly scoop up a bucket-ful of wrack and take it home to dry, spread out by the boiler. After several weeks, I have 400 grammes of dried weed, which I place in an open ceramic bowl in the fire. Orange flames from the sodium in the brine dance lazily as it burns, and afterwards I am left with just sixty grammes of a crispy grey ash. I pound this to a powder and stir into it a minimum of water to create a runny black sludge, which in turn I place in a funnel with a filter paper. A clear liquor trickles from the spout, rich with marine salts. Most of the solution will be sodium chloride, of course, but bromide and iodide should also be present. Seaweeds are efficient at concentrating these elements. The concentration of iodine in seawater is less than a hundred parts per billion, but in seaweed it can be several thousand parts per *million*, a hundred thousand times greater. I allow the filtrate to stand for a few days, during which time an impressive quantity of white salt crystallizes from the solution.

Now it is time to attempt the conversion of the colourless iodide into the gaudy hues of the pure element. Like Courtois, I add a splash of sulphuric acid and follow it with a good quantity of hydrogen peroxide (not terrorist grade, but fairly strong) which should oxidize the acidified iodide to iodine. I shake the mixture to speed it on its way and see the liquid begin to colour. Pale yellow darkens through shades of saffron and settles after a few minutes to the colour of stewed tea. I am truly amazed. I have never attempted the experiment before and have been entirely careless in collecting my raw material, but I have my iodine. Or nearly – this rich brown is due to iodine mixed with iodide salts. I still want to see the brilliant violet vapour that astonished Courtois.

I decant the brown liquid and shake it up again with carbon

tetrachloride. This sweet-smelling but unlovely chemical – carcinogenic and ozone-depleting – is practically unobtainable these days, but I have found some in my father's comprehensive selection of dodgy solvents. It does not mix with water but preferentially dissolves the iodine. In this very different solvent, I see for the first time the characteristic colour. Violet is the right word: it is far beyond mauve in intensity, yet lacks the sinister depth of a purple. I say a quick mea culpa for the sake of the ozone layer and allow the carbon tetrachloride to evaporate, leaving behind a black film on the glass. These are tiny iodine crystals. They emanate a faint pungent smell, similar to but less acrid than chlorine, not entirely unpleasant, the kind of smell that we now think of as medicinal, retrospectively applying our cultural knowledge that the halogens are used as disinfectants. I apply gentle heat to the crystals, and watch as the first pink wraiths begin to rise up in the test tube. Soon the solid is gone, and all that remains is an intensely coloured swirling vapour, which recondenses on the cooler parts of the tube – the same pure element, its atoms reconfigured in new black crystals. When Johann Wolfgang von Goethe performed the same experiment for the amusement of some house guests in 1822, he delighted in the support it gave to his influential theory of colours, which held that reds and yellows were related to white while the 'cool' colours at the violet end of the spectrum were derived from black.

## Slow Fire

If a person today knows only one chemical formula, it is sure to be $H_2O$, the formula for water, a compound comprising two parts of the element hydrogen to one of the element oxygen. In the eighteenth century, however, neither H nor O was known, and water itself was still widely believed to be one of the irreducible elements of which all matter was composed.

Ever since Aristotle, water had seemed the most secure of the four elements. On the occasions when philosophers and alchemists thought to question the theory, it was fire (which needed to feed on other elements in order to sustain itself), or earth (which so obviously comprised many distinct substances), or air (which might be nothingness itself) that gave them trouble. Water at least tended to look and feel like water, and remained the element most clearly linked to its 'principles' or fundamental properties of being cold and moist. Yet water was a puzzle too. It might appear constant, but waters from different sources often tasted very different, ranging from strangely refreshing to quite undrinkable.

Modern science had reason to investigate the nature of this Aristotelian element more closely. In the growing cities, sanitation was non-existent and clean water always in short supply. Utopian fictions always include a bountiful supply of pure fresh water on their inventory of benefits. The principal river of Thomas More's *Utopia* (1516) is the Anyder, its name derived from the Greek for 'no water', just as More's coinage of 'utopia' means 'no place'. The strangely Thames-like tidal river is no use for supplying the city with drinking water, which More describes as being brought instead by elaborate means of channels and cisterns. Francis Bacon's *New Atlantis* (1624) goes a scientific step further and imagines the purification by osmosis of water into 'pools, of which some do strain fresh water out of salt, and others by art do turn fresh water into salt'.

Hazily, the generation of natural philosophers who came after the alchemists began to understand that the *quality* of water mattered to public health. What drove them was not only a sense that filth contaminating the water was a cause of illness, but also a belief that certain substances added to it might make it positively health-giving. Out of this work by turns would emerge science's understanding of acids and salts as well as the isolation of water's gaseous ingredients, hydrogen and oxygen.

In 1767, the thirty-four-year-old nonconformist minister Joseph Priestley returned from one of his regular long visits to London to settle in Leeds, the city of his birth, and moved into a house adjacent to a brewery. A man of huge intellectual curiosity, he had written biographical and scientific histories, published pamphlets critical of Britain's policy towards its American colonies and challenged congregations by preaching his unorthodox variety of Christian belief. However, inspired by meetings in London with Benjamin Franklin, Priestley now found his true métier in experimental science. Moving to Leeds, it was only natural that he should turn his attention to the constant bubbling of the recently identified 'fixed air', emanating from the beer mash next door.

Priestley made a systematic study of the properties of this gas, noting that it would extinguish a flame and cause the asphyxiation of animals, but that plants thrived in it. He convinced himself that the gas had a beneficial effect against ailments such as scurvy, which led him to consider whether a convenient means could be found to administer it. By sloshing water from glass to glass over a tub of the brewer's barley mash, he discovered that some fixed air would dissolve in the water and realized that he had his answer. Priestley devised a general means of making the effervescent drink – for those not blessed to have a brewery on their doorstep – and in 1772 published 'Directions for Impregnating Water with Fixed Air', based on reacting sulphuric acid with chalk and then bubbling the gas released through ordinary drinking water. He suggested that the fizzy liquid that resulted might have both therapeutic and military applications.

'Fixed air' was, of course, carbon dioxide. The Frenchman Gabriel Venel had earlier combined much the same ingredients, but expected people to drink the whole sludgy concoction. Priestley's was the first drinkable carbonated water, although he did not exploit the discovery, leaving this to Jacob Schweppe, the Swiss émigré who in 1792 established the London soda water business that still bears his name.

In France, meanwhile, a long-running national exercise was under way to gather information for a mineralogical atlas of French waters. Venel had contributed data from his analysis of the waters of Selters, or Seltz, on the French banks of the Rhine in 1755, the authentic source of seltzer water. The young Antoine Lavoisier, who would rise to become France's greatest chemist, also played a part in this project. His experience here laid the groundwork for his journey of discovery: that 'waters' were simply universal water combined with different salts; that these salts were characterized in turn by different combinations of metals and acids; and that these acids generally gained their corrosive properties from their incorporation of the as yet unknown element oxygen.

Like his English competitors, Priestley and later Humphry Davy, Lavoisier was educated in the humanities but soon realized that the questions to match his intellect were to be found in science. At first, however, he followed in his father's footsteps, studying the law and purchasing a royal concession to collect taxes. His highly profitable duties covered the prevention of smuggling of drink and tobacco and the collection of the notorious *gabelle*, or salt tax, that would later be one cause of the French Revolution. Meanwhile, his scientific acumen was directed towards assaying the minerals in natural water sources. The work gave ample scope for Lavoisier to refine the rigorous analytical techniques that would seal his reputation for dragging chemistry out of the era of alchemy. He invested some of the fortune he made as a 'tax farmer' in building the best instruments that he could. By accurately measuring the very slightly different density of various waters, he was then able to say how much salt they contained. But he did not much enjoy the routine of days spent in the sun and rain and nights billetted in cheap inns. He preferred the comforts of the laboratory and worked hard to earn them.

While Priestley was experimenting with carbon dioxide, Lavoisier, by now more happily situated in Paris and freshly elected to the French Academy of Sciences, turned his measurement skills to combustion reactions. He found that diamond, sulphur and phosphorus when burnt in air all gained weight if the gases produced were figured into the calculation. The same happened on the slower scale of metallic corrosion. In 1773, he gave an important paper to the Academy, properly recording for the first time that the transformation of copper and iron into verdigris and rust was also accompanied by a gain in weight. He explained these observations by suggesting that the substances must be absorbing something from the air.

In October the following year, Lavoisier and his fellow academicians hosted Joseph Priestley at a dinner in Paris and heard of Priestley's latest experiment in which he had heated mercuric oxide (known as red calx of mercury) to release 'a new sort of air', leaving behind only pure liquid mercury. The month before, Lavoisier had received correspondence from Carl Scheele in Sweden, who had done the same experiment a little earlier. Scheele was an exceptionally modest fellow who never sought academic recognition and only ever attended one meeting of the Royal Swedish Academy of Sciences. He left behind him no reliable portrait, so that even his statue in a Stockholm park is a Grecian fancy rather than a genuine likeness, and, worst of all, he did not hasten to publish his work. Priestley, meanwhile, was in a theoretical muddle as to what he had found. This left the way free for Lavoisier, who repeated the two men's work and performed further experiments of his own before, in 1777, naming the gas oxygen, meaning generator of acid.

Priestley's scientific interest was chiefly in the gases of the air, whereas Lavoisier's was in the waters. And, like most Swedish chemists, Scheele's focus was on the minerals of the earth. Converging on this vital element from each of the three states of matter, gas, liquid and solid, it is hardly surprising that the three scientists had trouble comparing notes. However, the clouds of confusion would eventually lift to reveal the ubiquitous importance of the element in all of nature. It is fair to attribute the discovery of oxygen gas to Scheele and Priestley, but it is Lavoisier who locked the newfound element into the rest of chemistry by proving its centrality to water, acids and salts.

Eleven years earlier, in 1766, Henry Cavendish, a man of Getty-like wealth and eccentricity, had discovered hydrogen, or 'inflammable air', by reacting metals with acid in his private laboratory in London. He amused himself thereafter by setting off explosions by sparking mixtures of the gas with air. The liquid that condensed from these explosions was simply water. In this way, Cavendish confirmed that water was not an element because it could be created from other fundamental ingredients, namely hydrogen and something in the air.

Lavoisier was able to repeat Cavendish's experiment on a lavish demonstration scale using pure hydrogen and what he now knew to be pure oxygen in the summer of 1783. Apparatus of the kind that he used is preserved at the Musée des Arts et Métiers in Paris. Even today the fine brass work and elegantly blown glassware suggest the precision of Lavoisier's method. Two huge gasometers containing the gases were first weighed, before the gases were allowed to mix in a vast glass bulb. Wires running into the bulb created a spark that ignited the hydrogen.

The only residue was a few grammes of water, which showed conclusively that water was comprised solely of these two gases. That same summer, the Montgolfier brothers went aloft in the first hot-air balloon. Lavoisier saw immediately that if ultra-light hydrogen gas could be made economically in bulk from water, there would be a demand for it in ballooning.

I remember performing the same demonstration at a chemical happening which I organized at my school. The event was billed as Explo '76. The hydrogen–oxygen reaction was not the most col-ourful or the smelliest item on the bill, but it certainly produced the loudest bang when I set off the balloon containing the two gases using a lighted taper lashed to the end of a long stick. Indeed, the sharpness of the report was our gauge of how accurately the gases had been introduced in the correct proportions of two to one. A fraction of a second later a fine mist hung in the silent space where the balloon had been. The Explo events, I later learnt, had continued for some twenty years after I left the school until they eventually became so grandiose – I heard tell of demonstrations so bombastic that they were staged no longer in a lecture theatre but in the school's obsolete and drained outdoor swimming pool – that they attracted the attention of the emergency services.

I have tried to reach this famous turning point in the history of chemistry without using the dread word 'phlogiston', a concept so tenacious during the eighteenth century and yet so mistaken and confusing that it still has the power to deter the amateur scientist. Phlogiston was the 'principle of fire' which Priestley and many others at the time mistakenly believed to have ma-terial existence. Phlogisticated air is therefore air where combustion has taken place, and dephlogisticated air, perversely, is air with the *potential* for combustion. The confusion arises because a presumed absence (of phlogiston) in fact turns out to be a presence (of the element oxygen).

The phlogiston theory explained what chemists observed very well, but provided no real understanding of the processes involved. One way to picture the confusion is to think of a moulded mask of a human face. Strongly lit from the side, you can clearly see the peak of the nose and the sockets of the eyes. But it is only by changing your perspective, or better still by reaching forward and touching the mask, that you find the light is coming not from the right as you had thought, but from the left, and you are in fact seeing the face from behind and not in front. Phlogiston was just such a reverse image, accurate to all appearances, and yet still fundamentally deceptive. It required Lavoisier's altered perspective to see things as they really were.

Although it correctly explained nothing, phlogiston was a stubborn theoretical concept. Even Lavoisier, a notable phlogiston sceptic even before his experiments with oxygen, used terms such as *air déphlogistiqué* as well as *air empiréal* and *air vital* alongside his new word oxygen until at least 1784. In an amusing pre-echo of our present obsession with anti-oxidant creams, Gustave Flaubert makes reference to a 'pommade antiphlogistique' in *Madame Bovary*, which is set a good fifty years after the theory ceased to have any scientific currency.

Lavoisier's work placed oxygen – rather than fire – at the centre of combustion and so of much of chemistry. In 1789, on the eve of the French Revolution, he published an *Elementary Treatise on Chemistry*. It included a comprehensive list of 'simple substances belonging to all the kingdoms of nature, which may be considered as the elements of bodies'. These were divided into four categories. The first included the gases, hydrogen, oxygen and nitrogen, as well as light and 'caloric', or heat. The second comprised six non-metallic substances that formed acids – carbon, sulphur, phosphorus and the unknown bases of muriatic (hydrochloric), fluoric and boracic acids. The third category listed seventeen 'oxydable' metals from antimony to zinc, and the fourth added five 'salifiable simple earthy substances' including lime and magnesia that Lavoisier correctly intuited to be concealing further new metal elements.

Lavoisier's textbook sold well. He had launched a chemical revolution; now came the political revolution. Lavoisier was clearly sympathetic to the *ancien régime*, although he rejected Louis XVI's last-ditch invitation to become his minister of finance in 1791, claiming that to do so would jeopardize the 'ideal of balance' that he sought to bring to economics and politics as much as to chemistry. Across the Channel, meanwhile, Priestley threw a party to celebrate the anniversary of the fall of the Bastille, and later that day a royalist mob destroyed his home. Lavoisier was to suffer an even worse fate at the hands of the Jacobins – on 5 May 1794, he went to the guillotine, hated as a tax collector and ignored for his science.

It is possible that, if the concurrent discoveries of oxygen in air and water had not been made when they were, we would not now accord this element the importance that we do. The chemical revolution would have been postponed, perhaps not triggered until Alessandro Volta made the first battery in 1800 using electrodes of copper and zinc. Our perception of chemistry would then arise less from the doings of one ubiquitous, hyperactive element – gaseous but material nonetheless – and more from the fleeting exchange of incorporeal electrical charges between chemical bodies, and we would now be without 'the excessive domination of oxygen in doctrine and nomenclature'.

But oxygen did move to the chemical centre, and in due course also came to acquire a far broader symbolic role in our language. This did not happen immediately, as it did, for example, with electricity. The romantic writers famously saw the dramatic and metaphoric potential of galvanism, Mary Shelley's

*Frankenstein* being only the most celebrated work inspired by the new understanding of electricity. But they also took inspiration from the new chemistry. Where Shakespeare had to make do with 'sweet air' and 'summer's ripening breath', the poets of the nineteenth century could sample the concentrated essence of air and life and consider whether to add it to their lexicons. Coleridge attended Davy's lectures – he came in order, as he said, 'to increase my stock of metaphors' – and observed on one occasion how ether 'burns bright indeed in the atmosphere, but o! how brightly whitely vividly beautiful in Oxygen gas'. Another time, he noted how oxygen and hydrogen could be prised from water with electricity. Though intensely aware of the discovery of oxygen and of its role in life, the romantics did not put it in their poetry, however. Poems such as Percy Shelley's 'Ode to the West Wind' and 'To a Skylark' burst with life-giving air and water and the blues and greens they occasion in nature but do not mention oxygen by name. Perhaps they feared their readers were not so well acquainted with the latest science as they. More likely, they simply rejected the word as lyrically unfit, a polysyllable that paradoxically seemed to choke the flow of breath. Much later, Roger McGough got round the problem by using the element's smoke-ring of a chemical symbol rather than its name in his poem, 'Oxygen', whose last line represents a person's final breaths by a fading sequence of eight 'o's.

How then did oxygen come to gain currency as a metaphor for 'essence', so that we immediately understand, for example, the Victorian poet Francis Thompson's writing of Shelley how 'The dimmest-sparked chip of a conception blazes and scintillates in the subtle oxygen of his mind', or the vow of Margaret Thatcher – a one-time chemist, of course – to deny terrorists 'the oxygen of publicity'?

The answer may lie in the spread of oxygen therapy during the nineteenth century, which introduced the gaseous element to the public for the first time. Understood as necessary to support life, oxygen was now the gas of choice for use against all manner of ills. It could be readily made by heating saltpetre and was observed to produce feelings of 'comfortable heat' in the lungs and limbs. Oxygen treatment could relieve diseases that led to breathing difficulties such as phthisis (pulmonary tuberculosis), although the relief only lasted as long as the gas. Against many other ailments, oxygen had no obvious effect, but this of course was no barrier to those promoting the curative powers of 'vital air'. Early enthusiasm soon waned amid accusations of quackery, but a new method of producing oxygen from the air and storing it under pressure in easily transported cylinders led to a resurgence of interest in the middle years of the century. With very little proper medical investigation of the treatment, oxygen therapy was used largely indiscriminately and continued to be challenged by sceptics. 'A question is frequently asked, "Is Oxygen Gas Inhalation dangerous?" The reply is *decidedly, not at all so*; it can be used without any possible risk of harm, and always with a real hope of doing good,' ran one defensive advertisement in 1870.

Medical respectability came to oxygen therapy after the First World War, when the distinguished physiologist John Scott Haldane showed its beneficial effect on soldiers suffering the chronic effects of poison gas. Haldane was a notorious self-experimenter. He carefully exposed himself and compliant colleagues to various unpleasant gases in a sealed chamber known as the 'coffin' and noted their effects on body and mind. He climbed Pikes Peak in Colorado in order to breathe for himself the thin air at 14,000 feet. His major scientific contribution was to understand the role of haemoglobin in regulating breathing, but he also made a number of helpful innovations, introducing the decompression routine for divers and the miner's canary to warn of low oxygen levels underground.

The legacy of his work is seen in the now familiar terminology of oxygen masks and oxygen tents. Meanwhile, commercial products such as Oxydol soap also began to trade on the health-giving and cleansing properties of oxygen. Each box of Radox bath salts once explained its brand name as a contraction of the largely meaningless phrase: 'radiates oxygen'. The restorative promise of the gas lives on in the lately fashionable oxygen bars of Tokyo and Beijing, where for a fee one can breathe a purer air.

Once it was understood that it was not an element in its own right, ozone too – comprising three atoms of oxygen bonded in a triangle rather than the hand-holding pair of atoms in the oxygen we breathe – began to be marketed as what, in essence, it was, a more intense form of oxygen. It was called 'electric oxygen', a reflection of the means of its manufacture as well as an exciting piece of branding, and used to purify drinking water, remove odours and generally imbue all it touched with a healthy vigour. One bottled water bore the strapline 'OZONE IS LIFE', while long before the 'oxygen of publicity' there was (in John Dos Passos's *The Big Money*) the 'ozone of revolt'.

Recently, though, we have become inclined to view oxygen as the destroyer, not the supporter, of life. Following his experiments in which he observed mice thriving in oxygen and the increased rate at which candles burnt down, Priestley in his *Experiments and Observations on Different Kinds of Air* (1776) brilliantly foresaw that any creature given too much oxygen might 'live out too fast and the animal powers be too soon exhausted in this pure kind of air'. One of Priestley's fellow members of the Lunar Society, Erasmus Darwin, wrote in his poem 'The Botanic Garden' of oxygen as 'Air's pure essence' that nurtures plants and feeds the beating heart, but also as 'soft combustion'.

This flameless fire corrupts all that it touches. It is this ubiquitous, constant and inescapable reaction that has positioned oxygen centre-stage. It is why we classify many important chemical processes either as oxidations or their reverse, reductions. Oxidation does not always require oxygen itself. It can be accomplished by other chemical oxidizing agents such as chlorine or by the application of energy such as via ultraviolet light. Photosynthesis in plants uses light from the sun to promote both oxidation and reduction. The main reactions of photosynthesis convert carbon dioxide into glucose. But in another part of the forest, as it were, light oxidizes water (using manganese as a catalyst) to release oxygen, daily repeating in each green leaf the experiments of Scheele and Lavoisier. Oxygen is merely the waste product of these processes, a corrosive gas that would destroy animal life if animals had not evolved in tandem with the increasing levels of oxygen in the earth's atmosphere.

When they undergo chemical combination, the elements are said to exhibit different oxidation states. Often each is associated with a characteristic colour, such as the ferrous green and ferric brown of the salts of iron. But when iron rusts we are more likely to note the corrosion of time than to see the rich-hued beauty that Ruskin saw. Oxygen, 'that insinuating vamp', as another writer terms it, is the element that ruins others, crusting their pure surface with a layer of chaos and decay.

What is not yet oxidized is potentially so. The carbon in the wood of the trees is tomorrow's carbon dioxide. The rusting hulk is yesterday's ironclad battleship. Civilization, it is immediately apparent, is simply organized resistance to oxidation. We are able to stem the tide in some places, and even reverse it in a few by various desperate measures – wresting metals from their ores, planting forests, extinguishing fires – but never for long. Oxidation betrays the march of time and the inevitable triumph of entropy. The gas gives life and in doing so brings death closer. Oxygen is, according to a recent book on the element, 'the single most important cause of ageing and age-related disease'. Some of the damage comes from reactive chemicals produced as intermediates during normal respiration – not oxygen molecules, but the short-lived species containing unpaired oxygen atoms known as free radicals – that find themselves at a loose end, as it were, and able to wreak biochemical havoc. One of the most meaningful measures of ageing is to look at the extent of damage to biological cells due to this oxidation, the scientific equivalent of counting crow's feet or liver spots.

As I am writing this in June 2009, I hear that the singer Michael Jackson is dead at the age of fifty. Is it possible that the oxygen tent he reportedly slept under accelerated his life and brought forward his death, as Priestley observed and feared? Immediately, there is talk that his body will be preserved, in his trademark moonwalk pose, by 'plastination' using special resins, and displayed in the space where he had been planning to give a comeback concert: London's $O_2$ Arena.

## *Our Lady of Radium*

Every now and again, an element which even most scientists may never see will nevertheless escape the confines of the laboratory to achieve a kind of fame or notoriety in the wider world. It happened, as we have seen, with plutonium after the dropping of the atomic bomb. But it happened first with radium. An element – an explosively reactive as well as radioactive metal – of which no ordinary mortal had the slightest practical experience suddenly burst upon the world, was seized upon as a miraculous talisman, sought after and fought over, adopted for place names and product brands, and then was equally dramatically dropped a few decades later like a hot brick.

The central figure in the radium story – and one of the reasons why it became such a phenomenon – is Marie Curie. She was born Maria Skłodowska near Warsaw in 1867, but, excluded from university in Poland, emigrated to Paris to complete her education. She felt liberated in Paris, and still more so at the Sorbonne, where she was free to find her own direction without the stifling supervision she had known in her Polish *gymnasium*. Unusually, she studied both chemistry and physics; she would go on to win the Nobel Prize in both fields, an achievement still unequalled by any woman or man. Marie would have returned to Poland to follow her parents into teaching but then, as she prepared for her graduation exams, she met Pierre Curie; they married quietly the following year, 1895.

The next decade until Pierre's death at the age of forty-six, crushed under the wheels and hooves of a passing horse-drawn wagon, marked a scientific partnership of rare harmony and productivity. With Pierre's encouragement and space in his laboratory, Marie decided to investigate the spontaneous emission of X-ray-like energy – a newly reported effect that she termed 'radioactivity' – identified from samples of the uranium ore known as pitchblende. Her chief tool was a quartz device invented by Pierre some years earlier, which exploited the property that some crystals possess of emitting an electrical charge in response to pressure exerted upon them. This meter was capable of detecting the very small electric currents associated with radioactive decay processes. Marie found that radioactivity was a phenomenon intrinsic to particular substances, and not the product of some kind of interaction with other matter or energy as many people then thought. During the course of her measurements, she also found that some uranium ores were more radioactive than others, and that some – bizarrely – were even more radioactive than pure uranium metal. This could only mean that the ore must contain an unknown, highly radioactive material.

This caught Pierre's interest and, dropping his own research, he and Marie hastily began to pulverize a handful of pitchblende and then dissolve it using chemicals that would enable them progressively to isolate the most radioactive components. Over a period of two months, they gradually obtained a product 300 times more radioactive than uranium. They noticed that some of the radioactivity was linked to barium in the sample and some to the element bismuth. Three weeks later, they were convinced that a new element must be mimicking the chemistry of the bismuth, which is not naturally radioactive. The Curies had already chosen the name polonium after Marie's beloved homeland – she had once dressed as 'Polonia' at a gathering of expatriates in Paris – and on 13 July 1898 Pierre was able to write in the laboratory notebook the letters 'Po'. But their inability yet to separate the element from bismuth was a source of frustration especially to Marie. She wanted to hold polonium in her hand.

Meanwhile, the couple continued to chase down the radioactive species linked to barium using a new sample of pitchblende. They succeeded just before Christmas, this time obtaining unequivocal evidence for the existence of another new element, even more radioactive than polonium, to which they gave the name radium. The salts of barium and radium are more soluble than those of bismuth and polonium. It made sense to try to isolate radium by repeatedly boiling up salt solutions and then cooling them slowly so that pure radium chloride, which was marginally less soluble than barium chloride, crystallized out first. Marie set out to tackle this immense challenge in 1899. She acquired ten tonnes of pitchblende residue, already more radioactive than the basic ore. It came in sacks of brown dust mixed with pine needles. Processing the material in twenty-kilogramme batches, she turned the primitive 'hangar' of a laboratory into a factory with cauldrons of boiling radioactive liquor in various stages of preparation. The work was physically exhausting, but there was always the exhilaration of the chase. At last, by 1902, she had tangible evidence of the new element, a tenth of a gramme of pure radium chloride.

What does a chemist feel when she discovers an element? The sensations are often dissipated by lengthy effort, but there are moments of intense pleasure. With their two elements and their two Nobel Prizes, the Curies experienced more of these moments than most scientists ever do. Certainly, they were not enamoured of the official ballyhoo that came with their success. Attending even their own award ceremonies was never their highest priority, understandably so in Marie's case when the awards were sometimes grudgingly given – she had not at first been included on the Nobel nomination papers when Pierre was put forward with Henri Becquerel, the discoverer of uranium radiation. And the publicity that ensued was simply a nuisance.

But the material aspect of the discoveries thrilled them. Suspicion had quickly turned to conviction that the pitchblende was hiding something. Before long, they knew they were looking for new elements – and they had the names ready. Their scientific papers laid claim to the discoveries with a seemly boldness: they proposed these names without apology, but they were also generous in acknowledging the contribution of others. Marie in particular felt proud of 'our new metals', and was frustrated to know that radium and polonium existed without being able to take physical possession of them. They had hoped to see coloured salts, but delighted in the light that shone unexpectedly from the impure material. Sometimes after dinner they would sneak back to the laboratory to see the samples glowing in their places, a sight that never failed to stir them with 'new emotion and enchantment'.

How did radium, this rare, peculiar and intractable element, come to public notice? In the first place, of course, it did so because of the Nobel Prize. The seven awards made in physics, chemistry and medicine during the first two years of the prizes received little attention. But this changed dramatically with the first award made to a woman, and to a married couple, which handed the media material for all manner of romantic fantasies. The strange properties of radium – its luminous blue glow and its mysterious, invisible radioactivity – added spice to the mix. Marie Curie was beatified as 'Our Lady of radium', yet was also beginning to suffer from what was not yet known as radiation sickness.

George Bernard Shaw gauged the public excitement with satirical accuracy, but was too quick to deny that there might be real substance to it. Radium, he wrote in the introduction to his play *The Doctor's Dilemma*, 'has excited our credulity precisely as the apparitions at Lourdes excited the credulity of Roman Catholics'. For radium, whose ability to damage the skin had been noticed from the first, was now found to be miraculously effective in the treatment of cancers. This discovery at once launched an industry and a folklore. By 1904, there was a large brick factory on the banks of the Marne outside Paris making radium salts on a scaled-up version of the process used by Curie. Others quickly followed. Radium, destroyer of tumours, was too good to leave at that, and was quickly and indiscriminately exploited as a 'therapy' for ailments of the blood, the bones and the nerves.

Scientists rushed to experiment with the new element. William Ramsay bought a sample from a London chemical supplier and took it back to his laboratory to confirm that it was genuine. He put a little of the sample on a wire and put the wire in a flame. The red colour confirmed that it was pure radium, uncontaminated by barium, which would have turned the flame green, but the radioactive vapour that Ramsay unwittingly released into the laboratory as a result of the test rendered it useless thereafter for experiments in radioactivity.

Visitors flocked to the mountains where radium was naturally abundant – the Erzgebirge. These, the famous 'ore mountains' of Bohemia, were already known as the most prolific metal-producing region in Europe. The mines were reopened in 770 CE after the fall of the Roman Empire by Charlemagne, who brought in prisoners from Saxony – long celebrated for its miners – to obtain gold, silver and lead. Later, uranium and cobalt were mined as well, for coloured glasses and ceramics.

Joachimstal (now known as Jáchymov in the Czech Republic) became a centre of the tourist boom. In 1912, the Radium Palace Hotel, a massive neoclassical confection clinging to the wooded mountainside, opened its doors to offer radioactive spa treatments. The waters contained low concentrations of dissolved radium and gained a slight effervescence from its radioactive decay into radon gas. (Joachimstal has other elemental connections, too: in the sixteenth century, the first silver dollar coins, or *Joachimsthaler*, were minted here, and it was where Agricola wrote his metallurgical masterpiece, *De Re Metallica*.)*

The Radium Palace Hotel has recently reopened, promising treatments based on 'the healing effect of radon-rich waters that flow deep below the surface of the Earth'. If you're feeling flush you can book into the Madame Curie apartment. Not far away, another spa town still rejoices in the name Radiumbad. Radon water healing galleries were also widespread in the United States, where there were once settlements named Radium in seven states. There are still towns called Radium Springs in Georgia, Wyoming and New Mexico.

Spas have always been places of elemental renewal. The

---

* *De Re Metallica* wasn't successfully translated into English until 1912, the job finally done by the mining engineer and future American president Herbert Hoover and his wife, Lou Henry, a Latin scholar. It seems fitting to be able to report that in 2014 the US Mint will issue a Herbert Hoover dollar coin.

Romans came to Bath for the sulphurous waters. Bad Suderode in the Harz Mountains of Germany is the place for calcium, Buxton for magnesium, while Marienbad will spritz you with carbonated waters. Other waters are oxygenated or iodized. It seems only fair that this custom should keep abreast of chemical advance, and that the newfound elements radon and radium should also have their day.

Those who did not take radium at the source found radium brought to them. Radium was demonstrated at parties. People played radium roulette and went to radium dances. The 'Radium Models' posed in luminous costumes. Radium was popularized in cartoons and, above all, hailed as an all-purpose miracle cure. Radium was added to products of all kinds, especially those supposed to offer therapeutic benefit. The word appeared on many other items as a fashionable brand name. There was Radium butter, Radium cigars, Radium beer, Radium chocolate and Radium toothpaste, Radium condoms, Radium suppositories and Radium contraceptive jelly.

The public was before long familiar enough with the bizarre properties of radium that they were an effective means of enhancing almost any manufacturer's claims. Aurora Radium Fertilizer was sold with the promise that it 'heats the soil'. Radium was put in chicken feed in the hope that the eggs might be self-incubating, if not actually self-cooking. Oradium wool for babies was 'endowed with a physico-chemical treatment of remarkable power: radio-activity': 'Everybody knows the extraordinary effects of organic stimulation of cellular excitation passed on by radium . . . Wool so treated combines the standard advantages of the textile with undeniable hygienic value. To knit Baby's layette, children's woollen garments, your underclothes and your pullover, use LAINE ORADIUM.'

The Curie name was often invoked to endorse these remedies, illicitly in many cases. Curie Hair Tonic was said to restore hair growth and colour, for example. This commercial licence can be excused to some degree as the Curies' own Radium Institute would give its imprimatur to products where they genuinely contained a source of radium emanation. This was done out of scientific probity – a stamp 'du Laboratoire Curie de Paris' would discreetly guarantee that a preparation contained, for example, '5 millimicrogrammes de Radium élément pour 1 gramme de Crème'. The Radium Institute was also enlisted to brand chromium-plated bath-side dispensers of radiation. These *emanateurs* or 'fountains' bubbled radon gas from a decaying radium source along a rubber tube into the bathwater; they were also used to add radioactive fizz to drinks. They are now highly collectible.

The aura of an elixir is most evident in the illustrated boys' adventure books that made the element central to their quest. They positioned radium as an exotic material to be plundered from far-off lands with much derring-do. The splendid cover of one of these books, *La Course du Radium* (better translated, by the way, as *The Dash for Radium*, not *The Course of Radium Treatment*), shows tribal horsemen galloping through the desert too late to catch our hero making his getaway in a biplane. This was all pure fancy. For most practical purposes, the only sources of prepared radium were in the two most genteel and sophisticated cities of Europe, Paris and the Curies' laboratory, and Vienna with its rival Radium Research Institute.

It was abundantly clear by the 1930s that radium was a serious danger to health. The case of the New Jersey 'radium girls', who painted the dials on luminous watches, had seen to that. In 1925, one of these women sued her employer, the US Radium Corporation, for damage to her health. She and her colleagues were in the habit of using their lips to put a fine point on the brushes they used. In the end, at least fifteen workers died suffering extreme symptoms of anaemia and decay of the tissue in the jaw. Marie Curie was aware of the deaths of several French engineers who had been involved in preparing therapeutic radium sources, although there were at this stage none at her own institute, a fact that she put down to superior safety precautions, which were indeed remarkably thorough for the time. But very soon, a number of Curie's colleagues began to succumb to radiation sickness.

Despite the increasingly recognized danger, radium's popularity as a brand remained undimmed. French pharmacies sold 'Tho-Radia' eau de cologne, powder, cream soap and lipstick 'according to the formula of Dr Alfred Curie' – the doctor in point being either an impostor or a figment of the manufacturer's imagination as there was nobody of that name in the Curie family. Tho-Radia cosmetics, advertised as 'scientific beauty products' and promoted by one Jacqueline Donny, who was Miss France in 1948 and Miss Europe in 1949, may or may not ever have contained thorium and radium – the Curie Institute found none when they tested them. Many other products plainly had no business incorporating radium at all. Nevertheless, Radium razors traded on it, promising that they had 'the scientific edge'. A brand of 'parfum atomique' depicted a bottle labelled 'Atome 58' with a glowing halo around it, no matter that the element with the atomic number fifty-eight is harmless cerium. The last few brands failed as public opposition to nuclear weapons and nuclear power grew stronger in the 1960s. Radium itself is now restricted to use in radiological clinics.

The room where Curie discovered polonium and radium, which she later recalled as 'a clapboard hut with asphalt floor and glass roof giving incomplete protection against the rain', no longer exists. Science does not sanctify the spaces where breakthroughs are made, only the breakthroughs themselves, and occasionally those who make them. The Curie couple themselves embodied the extremes of the attitudes scientists may take towards their achievements. Marie admired Pierre's attitude that it did not matter who made a discovery so long as it was made, but could not share it, always feeling more possessive about her scientific achievements. Had it survived, the laboratory would have served as a reminder that discovery does not require comfortable surroundings, merely the right equipment at the right time, in this case the pitchblende and Pierre's sensitive quartz balance. Marie Curie wrote of that time that she and Pierre had been 'living with a sole preoccupation, as if in a dream'.

In 1914, eight years after Pierre's death, Marie Curie at last moved to more adequate quarters, a cluster of new buildings comprising the Radium Institute and, across a small garden, the Pasteur Institute. French windows in Marie's laboratory opened on to a small garden between the two buildings, symbolizing the closeness of chemistry and biology to nature and each other. Marie occupied this office until her death in 1934, whereupon she was succeeded as director by André-Louis Debierne, who discovered another element in pitchblende, actinium. Later, Marie's daughter Irène and her husband Frédéric Joliot-Curie took over the helm. In 1958, the building was closed because it was too saturated with radiation to do anything else with.

In 1995, however, it reopened as the Curie Museum. I meet the museum's coordinator, Marité Amrani, who has a refreshingly unParisian enthusiasm for her work. She shows me examples of radium-branded products before leading me into the rooms where Marie Curie did most of her work. She assures me that the place has been pronounced safe, but the dishevelled state of the cupboards and the antique bottles of chemicals left on the shelves make me wonder. I examine a sample of pitchblende, a dull-grey rock with hints of pinkish sparkle, and wonder what emanations it is still giving forth. Displayed on the wall is a page of Marie Curie's notebook, and alongside it a blackened radiograph of the same page betraying the heavy contamination. Her lab coat – black with white polka dots – betrays a hint of Parisian chic. In a corner is the mahogany box that once contained the gramme of radium that Marie accepted as a gift from the women of America who had raised the $100,000 necessary to acquire it. Inside the box is a solid cylinder of lead the size of a Stilton cheese with a small well dropped into the centre to house the radioactive source. I try and fail to lift it – 'It weighs forty-three kilogrammes,' Marité tells me. 'And today you would use much more lead.'

One of Marie Curie's greatest legacies is the peer effect that she created. 'She welcomed many women into the laboratory here,' says Marité. 'If someone was made for science, she would encourage them.' Marie's daughter Irène was her most obvious protegée, who went on to win her own Nobel Prize jointly with her husband – the second woman after her mother in both distinctions – in 1935. Another was Marguerite Perey, who discovered her own new element, francium, in 1939. Perey rose, like the restaurant *plongeur* who becomes the chef, from test-tube washer to be first Marie's personal preparatory assistant and then a fine scientist in her own right. Her discovery, made on the eve of the Second World War, met with none of the fuss that so irritated the Curies. Perey had first proposed the name catium and the symbol Cm for the element preceding radium in the periodic table (because of its predicted likelihood of forming highly reactive positive ions, or cations), but by the time that new element names next came up for official consideration, in 1947, a flurry of other radioactive elements had been discovered as a consequence of the Manhattan Project. One of these new elements had a better claim to the symbol Cm: curium. Perey accepted her second choice of name, francium. In 1962, she became the first woman to be elected to the French Academy of Sciences, which had chauvinistically excluded both Marie and Irène. Perhaps she named her element wisely in the end.

On my return from Paris, I got off the Eurostar and made my way to my parents' house, which I was using as a London way-station. I went to wipe from my black shoes the chalk dust that had settled upon them from walking through the Paris parks, and was astonished to find along with the tins of Meltonian polish a rectangular carton of black leather dye branded 'radium' in bold 1960s lettering.

## Nightglow of Dystopia

Gas was the principal means of lighting city streets and town houses from the middle of the nineteenth century. Its hissing white light was excitedly evoked in its prime and was still missed long after its demise. By the time that incandescent electric light was taking over around the turn of the century, the mere image of gaslight was sufficient to deliver a powerful nostalgic kick. In the famous German wartime song 'Lili Marleen', written in 1915, Lili is simply presented standing underneath a lamp-post (*Laterne*). By the time of the Second World War, however, when the song enjoyed renewed popularity, the English translation has her repackaged as 'Lily of the lamplight'; the allure is as much for a bygone age of innocence as for the *femme fatale*.

The wonder of artificial illumination naturally finds its way into descriptions of the urban world. Yet its light is not simply light. It radiates, illuminates and leaves shadows according to its kind, and in doing so establishes moods to which writers have been more or less sensitive. Dark deeds might be done by its beams, but gaslight itself – understandably since it was the first public lighting – was an innocent marvel. Even in novels beset with shadows, such as Joseph Conrad's *The Secret Agent*, gaslight comes out of it well. Indeed, Conrad is at pains to point out that its light is completely neutral. At one point, it catches the antiheroine Winnie Verloc's cheeks glowing 'with an orange hue'. This orange is no effect of the illumination; it is the composite of a red blush seen through her bilious yellow complexion. The white of the gaslight shows things in their true colours.

Writers have greeted the modern innovation of sodium street lighting differently. Like gas, the incandescent lamps that sodium would duly replace shone a generous white light, combining light of many colours, created by the flow of an electric current through a metal filament. Sodium on the other hand shines light of a single wavelength – 589 nanometres. When light from a sodium discharge strikes a colourful object all we see is the fraction of that 589-nanometre light that is reflected and no other colour. This monochrome wash is deceptive, not truth-telling; it soaks everything in a nicotinic glare such that it is no longer possible to perceive colour accurately.

The first sodium lights were installed in the streets adjacent to the lighting manufacturers themselves, Osram in Berlin and Philips near Maastricht in the Netherlands. Purley Way, near the Philips factory in Croydon, was the chosen British test site in 1932. As sodium street lamps became more commonplace after the Second World War, their staining light came to the attention of writers seeking to convey a sinister city atmosphere. In *Nausea*, Jean-Paul Sartre's alter ego, the young writer Roquentin, is tormented by his pointless existence, the 'Nausea' of the title; at one point he crosses the street to the pavement opposite, drawn by 'a solitary gas lamp, like a lighthouse', and is astonished to find that 'The Nausea has stayed over there, in the yellow light'. The poet John Betjeman, while fond of the Metroland which it illumined, reviled the 'yellow vomit' thrown out by the new concrete 'gallows overhead'. A generation later, J. M. Coetzee makes this idea work harder in his novel *Age of Iron*, set in apartheid-era South Africa. Coetzee's narrator, Mrs Curren, a retired professor who is dying of cancer, is being driven with her maid into one of the townships, where they will discover the body of the maid's son murdered by the police. The car splashes 'through pools on the uneven road . . . under the sick orange of the streetlights'. The light is a metaphor both for her cancer and for the cancer that is destroying the country. Anthony Burgess and J. G. Ballard also bathe their dystopian visions in sodium light. The element was surely a worn cliché by the time that Will Self, in *The Book of Dave*, has his eponymous London taxi driver eyeing up potential fares loitering 'frowsty under the sodium lamps'.

Joseph O'Neill manages to refresh the image in his 2008 novel, *Netherland*. The central character is coming to terms with his wife's decision to leave him. Staring out from the balcony of his apartment in New York's Chelsea Hotel, he bitterly twists a metaphor of potential sunrise into a *Götterdämmerung* sunset:

a succession of cross-streets glowed as if each held a dawn. The tail lights, the coarse blaze of deserted office buildings, the lit storefronts, the orange fuzz of the street lanterns: all this garbage of light had been refined into a radiant atmosphere that rested in a low silver heap over Midtown and introduced to my mind the mad thought that the final twilight was upon New York.

The Reagan-era Three Californias trilogy by science-fiction writer Kim Stanley Robinson presents different scenarios for the golden state. The second novel in the series, *The Gold Coast*, depicts perhaps the most likely of these futures, neither post-nuclear nor ecotopian. Here, Robinson riffs more extendedly on the lights of Los Angeles and their elemental origins:

The great gridwork of light.

Tungsten, neon, sodium, mercury, halogen, xenon.

At groundlevel, square grids of orange sodium streetlights.

All kinds of things burn.

Mercury vapor lamps: blue crystals over the freeways, the condos, the parking lots.

Eyezapping xenon, glaring on the malls, the stadium, Disneyland.

Great halogen lighthouse beams from the airport, snapping around the night sky.

An ambulance light, pulsing red below.

Ceaseless succession, redgreenyellow, redgreenyellow.

Headlights and taillights, red and white blood cells, pushed through a
leukemic body of light.

There's a brake light in your brain.

A billion lights. (Ten million people.) How many kilowatts per hour?

Grid laid over grid, from the mountains to the sea. A billion lights.

Ah yes: Orange County.

On every continent, sodium is now the colour of the city at
night and the principal means of our knowing this element, its
lurid, unlovely light an inescapable feature of metropolitan life.
Even the manufacturers and authorities responsible for install-
ing them recognize that sodium lamps are no triumph of
aesthetics, but they are favoured nevertheless because they are
more energy-efficient than the alternatives. Attempts to change
over to whiter lights based on mixtures of other chemical
vapours have been thwarted by successive oil crises, and so we
go about our nocturnal lives under sodium's singular glare.

It is not the 589-nanometre colour that offends. In another
context, this can offer cheer, as when sea-salt tints the flames of

a driftwood fire. It is the foggy ubiquity of it. I confess I share the general distaste for this artificial illumination inflicted city-wide, though I have only happy memories of the single sodium lamp that shone from the other side of the street into my bedroom as a child. I can recall watching how it flickered with fresh-washed pink (due to neon added to activate the sodium at a lower voltage) when first shocked into action on damp autumn evenings before brightening and passing through red and orange on its way to the full radiance that meant I had no need of a night light. I had not read any dystopian novels then.

It was not its characteristic light that led chemists to the discovery of sodium, as was to be the case with various elements identified later. In 1801 Humphry Davy moved from Bristol to take up a position as director of the laboratory at the newly founded Royal Institution in London. He took with him his galvanic piles, the primitive batteries with which he had lately begun to experiment, and a hunch that the electricity they generated might be key to the discovery of 'the *true* elements' of substances.

At the Royal Institution, he built more powerful piles by interleaving dozens of square plates of copper and zinc in elongated boxes like Christmas packs of After Eight Mints. He

summarized his first experiments with the new apparatus in a prize lecture to the Royal Society in November 1806. It was a piece of work of such promise that it immediately secured his international reputation, including the award from Napoleon that provided the reason for his later trip to France. Having concluded an investigation of the electrolysis of pure water and various solutions by this method, Davy turned his attention to melted salts. The following October, he immersed the platinum wire electrode of his battery into molten potash and almost immediately managed to decompose the material and produce a highly reactive new metal. Davy 'danced about the room in ecstatic delight at the end of it', according to his cousin Edmund, who had been enlisted as an assistant. A few days later, Davy repeated the experiment with the corrosively alkaline caustic soda, or sodium hydroxide, in place of the potash, and the same thing happened – another new metal.

In November, he returned to the Royal Society to give the same prize lecture, a performance that would trump the achievement of the previous year. Davy described how 'a most intense light was exhibited at the negative wire, and a column of flame, which seemed to be owing to the developement of combustible matter, arose from the point of contact'. The metal obtained from the potash was liquid and looked like mercury, while that from the soda was silvery and solid. Both were dangerously reactive: 'the globules often burnt at the moment of their formation, and sometimes violently exploded and separated into smaller globules, which flew with great velocity through the air in a state of vivid combustion, producing a beautiful effect of continued jets of fire'. Davy announced that he had chosen the names potassium and sodium for the new elements. But were they metals? They were extraordinarily light. If it were not for the fact that they exploded on contact with water, they would easily float on its surface. He found they floated even on naphtha, a petroleum oil considerably less dense than water. He concluded that their exceptional lightness should not be considered as overruling their other properties, such as high electrical conductivity, which showed them to be indubitably metallic. Using his uniquely powerful electrolytic apparatus, Davy had just discovered the two most reactive metals known to science.

Chemists strongly suspected that other minerals would prove to contain further explosively reactive new metals that simply awaited a powerful enough force to prise them free. One of these minerals was lime, which Lavoisier had included in his list of 'simple substances' on this promise; another was magnesia, which Joseph Black in Edinburgh had shown to be chemically analogous to lime and therefore likely to be a compound of a closely related metal. Strontia and baryta were two more substances that had been obtained by Black's pupil Charles Hope, who had noted their coloured flames (red and green respectively) as indicating the presence of new elements. Davy proceeded to submit each of these so-called alkaline earths in turn to his electrolytic treatment, this time using an electrode of liquid mercury to capture the metals as they were released in an amalgam before they could burn away. Through the course of 1808, Davy succeeded in isolating, one after another, calcium, magnesium, strontium and barium.

Chemistry was not Davy's only talent. He was also a romantic poet of serious promise. Robert Southey, later Poet Laureate, included some of Davy's verse in the *Annual Anthology* that he edited, and admiringly called him 'the young chemist, the young everything'. Davy saw no contradiction between his science and his art, linking the study of nature with a love of the beautiful and the sublime. The first stanza of a poem he wrote at this time seems to incorporate images of the inflammable metals released so dramatically from unyielding minerals:

> Lo o'er the earth the kindling spirits pour
> The flames of life that bounteous Nature gives;
> The limpid dew becomes a rosy flower.
> The insensate dust awakes, and moves, and lives.

Two further members of the highly reactive group of elements known as the alkali metals were found, unlike Davy's sodium and potassium, by means that *did* depend upon the signature light of their salts. In 1859, Robert Bunsen and Gustav Kirchhoff in Heidelberg made a spectroscope, a kind of sophisticated prism that enables scientists to identify elements by separating out the colours they give in a flame (provided perhaps by one of Bunsen's famous burners) into characteristic lines like a barcode. Bunsen and Kirchhoff used their new gadget to make a systematic investigation of the dissolved ingredients of mineral waters in case an undiscovered element should lurk there. By chemically removing the obvious salts of soda and lime, and the less obvious strontia and magnesia, they were left with a solution of rarer salts, from which they then evaporated all the water. Placing the solid residue of this solution in a flame, Bunsen and Kirchhoff observed a new, blue, light, which could only be due to an undiscovered element. They named it caesium, after *caesius*, the Latin word for the colour of the sky. A few months later, they followed a similar procedure on a mineral sample from Saxony and saw dark-red lines of another new element: rubidium.

A fifth alkali metal, lithium, had been found some years before by more conventional methods (named therefore not for its light in a flame but after the earth – *lithos* in Greek – in which it was found). Now, thanks to spectroscopy, it seemed that these metals were everywhere. One morning Bunsen surprised his co-worker by announcing, 'Do you know where I found lithium? In tobacco ashes!' The element had previously been thought to be very rare.

The existence of these relatively uncommon, but hardly rare, elements, caesium, rubidium and lithium, had simply been obscured by the omnipresence of sodium. Sodium is by far the most abundant alkali metal in the salt of the earth, and its bright yellow light easily washes out other colours from a flame. When astronomers complain of light pollution, it is often sodium street lights they have in mind. Edwin Hubble escaped the glare of 'Orange county' by retreating to a mountaintop observatory north of Pasadena, where he recorded the motions of the galaxies that led to his discovery of the expanding universe. But it wasn't sodium that caused him difficulties. Potassium burns with a mauve flame which can sometimes be seen in a gunpowder explosion or when lighting a match. One night Hubble was excited to detect a potassium spectrum while he examined the galaxies through the world's most powerful telescope. But it soon became apparent that the reading must be false. Eventually Hubble realized that the equipment had picked up the light from the potassium in the match that he had used to light his pipe.

The makers of fireworks, unlike the suppliers of artists' paints or prepared foods, are under no obligation to declare the chemical content of their goods. To those with rudimentary knowledge, their names may suggest certain ingredients. My cheap Fifth of November box made broken-English promises of 'silver glittering', 'green diamonds fountain' and 'golden nuggets'. Probably magnesium, copper and sodium, I thought. But verification comes only when the skies are illuminated in the elements' signature shades.

Different yellows and oranges are created by sodium salts, powdered charcoal and iron filings, for example. Green has traditionally been made using copper salts, such as verdigris. Long before they knew about the other elements that could answer their wishes, pyrotechnists wanted to recreate the full spectrum of colours through their craft. The Chinese achieved something approaching the effect by using ribbons of coloured paper as filters through which the light of their exploding gunpowder could shine. As early as the mid eighteenth century, fireworks were advertised as offering proper rainbow colours. But in truth, it seems that the colours are brighter in description than they can ever have been in the fireworks themselves. Gold and silver were the predominant tones, obtained from various mixtures of powdered iron and the black sulphide ore of antimony, which sparkled orange and white respectively.

King George II attended one of the most elaborate displays of the age at Green Park in London in 1749, following the signing of the Treaty of Aix-la-Chapelle. Handel wrote a 'grand overture on warlike instruments', the piece we now know as his *Music for the Royal Fireworks*. However, Horace Walpole was

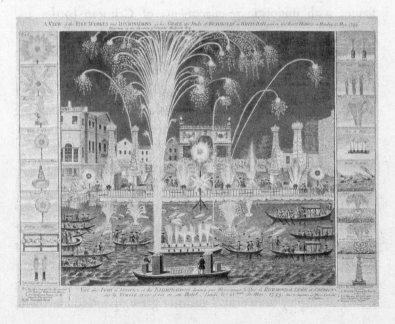

disappointed that the display itself was 'pitiful and ill conducted with no change of coloured fires and shapes . . . and lighted so slowly that scarce anybody had patience to wait for the finishing'. Even if things had gone to his satisfaction, the green of copper would have been the only colour he would have seen apart from the whites and yellows seen in all incandescent fire.

Charles Dickens's 1836 *Sketches by Boz* delights in 'red, blue, and parti-coloured light' at one display, while in *Pendennis* (1848) William Thackeray has the girl Fanny Bolton thrilling to fireworks of 'azure, and emerald, and vermilion!' Both descriptions imply an intensity of colour far beyond that achievable at the time, and bear witness more to the ever-wishful imagination of the fireworks spectator. Even when strontium and barium salts became available, the red and green colour they gave was often still feeble owing to the presence of impurities.

The early fireworks displays were abstract affairs, but in Victoria's reign a fashion developed for pictorial representations in flame, with jingoistic re-enactments of Crimean battles and Indian campaigns being especially popular. When there were fewer glorious victories to report, the trend reverted to displays in which there was less to distract from pure pyrotechnic artistry. However, public enthusiasm for fireworks almost died altogether when the novelty of gaslight led to an alternative fad for adorning major buildings with special illuminations at times of celebration.

These days, firework displays are relayed on television, the europium and zinc of the phosphor screen making a feeble imitation of sodium and barium in the night sky, and there are new fears for the pyrotechnist's art. Hidden among the bushes in a Cambridgeshire layby, I find an unmarked gate that opens to admit me to the redoubt of the Reverend Ron Lancaster, the managing director of Kimbolton Fireworks, Britain's last remaining maker of display fireworks. Lancaster grew up in Huddersfield, the historical centre of the British firework industry, and began making his own fireworks there during the Second World War. (These were the days when you could easily buy saltpetre and mix your own gunpowder.) He became a curate and later chaplain at Kimbolton School, where he taught the unusual combination of divinity and chemistry. The summer holidays provided ample opportunity for giving firework displays. In 1964, he built a laboratory to pursue his pyrotechnic experiments and finally set up the company.

For a man devoted to bringing joy as well as salvation into people's lives, I find the reverend in a gloomy mood. The industry cannot survive much longer, he fears. He rattles through a long list of obstacles: 'health and safety propaganda, supermarket BOGOFs, Chinese imports, bureaucracy'. One protestor wrote to Lancaster: wasn't he ashamed that his fireworks were filling the atmosphere with cadmium and mercury? 'I wrote back: look at the crematoria, and the mercury fillings and the exploding pacemakers, I said.' I can see he faces problems. Following a spate of 'stupid' accidents and vigorous consumer campaigning, fireworks retailers have been subjected to a tightening ratchet of restrictions – the noisiest bangers were outlawed, then fireworks with erratic flight, and other fireworks were muffled or tamed. Yet it is anti-social usage rather than the intrinsic danger of fireworks themselves that has driven the new legislation. Most of all, Lancaster regrets the side-effect of all this, which has been to initiate a trend away from back-garden fireworks towards large municipal events, leading to 'the control of big displays by people who hate fireworks'.

November the Fifth is no help, either. 'It's an awful day.' Lancaster believes Britain would be happier about fireworks if our annual excuse for letting them off did not fall in this dank month. But a kind of Dunkirk spirit means that we stubbornly tough it out each year without ever really enjoying the spectacle. 'Our phlegmatic approach has killed it. Go to Spain, and see how fireworks have to be part of every fiesta in every community.' By email, I poll a selection of friends in the United States, Israel, Russia, Italy, Spain . . . and indeed receive in reply a barrage of festive occasions when fireworks are let off.

Fortunately, perhaps, the Reverend Lancaster's passion is not running the business, but pyrotechnic research. I steer the conversation on to the problem of colours. Lancaster's first breakthrough came when he was offered a supply of titanium turnings from an aircraft machine shop. Although they are tricky to handle – they are very hard, which makes them sensitive to friction and hence liable to trigger an accidental ignition – he found a way to incorporate them safely in fireworks, where they burn to produce beautiful silver sparks. A century before, aluminium and magnesium had been introduced into fireworks to similar effect, but titanium is brighter and, moreover, immune to damp. For a time during the 1960s, its white sparkles became quite a fad.

One of Lancaster's goals was to create new incandescent colours intermediate between those made by the well-known chemical salts. One target was lime-green (barium and copper burn with more of a sea-green colour). Because he is dealing with dazzling light, the pyrotechnist's craft is more subtle even than that of the artist mixing paints, combining elements of chemistry, ballistics, optics and perception. In the case of lime-green, simply blending the green of copper or barium and the yellow of sodium was not the answer because each colour requires a different flame temperature. Adding magnalium (an alloy of magnesium and aluminium) enabled Lancaster to produce the component colours under greater control at a higher temperature, but this then required the addition of further chemicals to give them intensity.

The creation of a good orange light is likewise not simply a matter of blending the red of strontium, say, and the yellow of sodium. Lancaster discovered that, for some reason to do with human visual perception, a little green is also necessary to produce the desired effect. His eureka moment came at the local cinema as he watched the lights of the Wurlitzer merging from red to green, momentarily producing the colour he was after.

Blue has proved especially elusive. In Napoleonic France, Claude-Fortuné Ruggieri was the first to make systematic use of metal salts to produce coloured flames. These were used for military signalling as well as for public spectacles. He published many editions of his *Elémens de Pyrotechnie* through the first half of the nineteenth century, giving recipes for many colour compositions but never a blue. No commonly available metal or salt produces a strong blue emission – a blue demands more energy than is typically released from the electronic transitions of excited atoms that generate light. All sorts of substances were tried in the nineteenth century, from ivory to bismuth to zinc, but the best colour that could be managed was a cold white that only looked blue alongside some yellower light. Thackeray's 'azure' was pure exaggeration. Only later was it learnt that copper salts that naturally burnt with a green flame could be chemically modified to burn blue. Before modern regulations, manufacturers sometimes used the poisonous and unstable copper acetoarsenite, the pigment artists call Paris Green, for this purpose. More recently, it has been found that the effect can be produced by the less noxious expedient of burning copper in the presence of chlorine. For good measure, the pyrotechnist will also often trick the eye by sending the blue up together with some contrasting light to produce the illusion of a deeper hue.

I am given to understand that psychology matters as much as chemistry in creating the perfect fireworks display. Today's organized shows draw large audiences and consume massive amounts of ordnance. The professionalism is admirable, with each firework set off electronically often to the beat of accompanying music with a precision that would have caused Handel to marvel. But the Reverend Lancaster deplores even this development. 'The problem is that it all happens too quickly, because it is made to be continuous to fit with the music.' He makes a more subtle point: 'What you see, and what you thought you saw, depends very much on your viewpoint and the conditions.' A massive coordinated public display can still disappoint if the weather or the crowds so determine. All the quick-fire razzmatazz can be poor compensation for the cordoned remoteness from the action, whereas a small-scale spontaneous display – Lancaster recalls standing, drink in hand, with friends on the beach at Aldeburgh after the summer carnival, and letting off a few rockets at intervals over the sea – is more likely to be remembered.

And, as I find when November the Fifth rolls round, mild and dry enough, even a modest pack of fireworks is enough to occasion wonder. The colours, red and green, are scorchingly bright. Occasional white flashes produce a retinal burn against which showers of orange sparks from iron filings appear merely brown and hardly luminous. By some chemical or perceptual hocus pocus, one firework produces a quite deep indigo, more an absence than a presence, a momentary void of light in the sky. A simple catherine wheel my nine-year-old son interprets as a solar eclipse, as its bright disc first gathers pace, forcing the firelight centrifugally out to the rim to form a dazzling corona, before rematerializing as a luminous disc once more as it slows and finally dies. The reverend is right. There is more magic here on this muddy field edge, feeling the rain of gritty soot as each rocket goes up and savouring the sulphur fragrance in the misty air.

## Cocktails at the Pale Horse

In *The Pale Horse* by Agatha Christie, a string of murders is found to be attributable to poisoning by the element thallium.

Why did Christie choose such a recherché material when she had free rein of all the poisons known to man? How did she know about it?

Thallium was controversial from its first public appearance at the International Exhibition held at South Kensington in 1862, where it was the bone of contention in a sharp scientific dispute. Inspired by Bunsen and Kirchhoff's discovery of caesium, a young chemist called William Crookes at the Royal College of Chemistry acquired his own spectroscope – one of few in the land – and in 1861 began to turn it on his experiments. Investigating a particular mineral from the Harz mountains, from which he was hoping to obtain tellurium, he observed an unfamiliar line in the green region of the spectrum. 'Have you ever noticed a single bright green line, almost exactly as far from Na [sodium, yellow] on one side as Li [lithium, red] is on the other side. If not, I have got a new element,' he wrote to his collaborator. He had indeed got a new element, which he named thallium, after the Greek for the green shoots of new plants, for the discovery was made in the spring. (If thallium weren't so scarce and poisonous, it might do for Ron Lancaster's lime-green.) Crookes began scraping together enough of the element to display at the coming exhibition, hopeful that it might assist in his election to the Royal Society.

Meanwhile, Claude-Auguste Lamy, who was professor of science at the University of Lille in France, also isolated thallium, extracting it from residue lining the lead chambers of a sulphuric acid plant. In June 1862, he arrived in London carrying with him a fourteen-gramme ingot of the new metal, which he unveiled at the exhibition, declaring Crookes's black powder specimen to be no more than an impure sulphide. Crookes was peeved when the Frenchman was awarded an exhibition prize, and enlisted his friends in the scientific press, who loudly proclaimed him as the first British discoverer of an element since

Humphry Davy. Crookes duly obtained redress from the exhibition organizers and the following year gained the Royal Society fellowship he coveted.

In Agatha Christie's thriller, the shady goings-on that we are first made aware of revolve around an old inn, the Pale Horse, which is occupied by three 'witches' who are apparently prepared to arrange murders. A hit list is found. Those already found dead have succumbed to sicknesses displaying symptoms of such variety that it is initially supposed they must all have died of unrelated natural causes. However, Mark Easterbrook, the hero of the tale, has his suspicions aroused when he learns that one of the victims' hair was falling out. 'Thallium used to be used for depilation at one time – particularly for children with ringworm. Then it was found to be dangerous,' he explains. 'It's mainly used nowadays for rats, I believe.' It transpires that the coven is a smokescreen, the witches don't carry out killings to order, and the murders were perpetrated by the 'witness' who first implicated them, by replacing objects in his victims' homes with substitutes contaminated with thallium.

Christie clearly chose thallium in order to prolong the mystery. It is the sheer diversity of the victims' symptoms that has the book's characters and us mystified for 300 pages. How did Christie know about it? She tells us through the person of Easterbrook, who is conveniently asked: 'What put thallium into your head?' He replies: 'I read an article on thallium poisoning when I was in America. A lot of workers died one after the other. Their deaths were put down to astonishingly varied causes. Amongst them, if I remember rightly, were . . .' and he goes on to itemize twelve diagnosed causes of death and five symptoms (presumably so that we know Christie has done her homework).

*The Pale Horse* 'popularized' thallium, and is surely one reason why it was at first suspected as the poison used against the former Russian spy Alexander Litvinenko, who was assassinated in London in 2006. (The cause of death turned out to be the even more exotic radioactive polonium, although it is likely that the KGB did use thallium in poisoning another dissident, Nikolai Khokhlov, in 1957.)

In other cases, awareness of the dangers of thallium promoted by Christie's thriller may have helped to foil real-life killers. Reversing the usual presumption that murder fictions encourage copycat killings, *The Agatha Christie Companion* gives three instances where, it claims, 'the symptoms of thallium poisoning . . . were recognized, and lives saved, because of the quick thinking of individuals who just happened to have read *The Pale Horse*.' In one instance, a Latin American woman wrote to the author to say that she had identified the symptoms in a man who was being slowly poisoned by his wife. A year or two later, a nineteen-month-old Qatari girl was brought to the Hammersmith Hospital in London apparently dying of a mysterious disease. The doctors were baffled, but a nurse who had read *The Pale Horse* suggested treatment for thallium poisoning. The infant had ingested thallium used by her parents as insecticide.

The third and most alarming case occurred at the Hadlands photographic works at Bovingdon in Hertfordshire in 1971. Around seventy people were made ill by what became known as the 'Bovingdon bug', and two died. The workers suspected environmental pollution, but tests at the factory revealed nothing. At a meeting, the company doctor ruled out heavy-metal contamination, but one worker, Graham Young, interrupted: 'Do you not think the symptoms are consistent with thallium poisoning?' The forensic specialist brought in by Scotland Yard, meanwhile, remembered the symptoms described in *The Pale Horse*. When the police searched Young's flat they found large quantities of thallium, and in due course he was found guilty of the murders. After the trial, it emerged that he had been recently released from Broadmoor high-security psychiatric hospital, where he had been imprisoned nine years earlier for attempting to poison most of his family including the cat.

The authors of *The Agatha Christie Companion* don't comment on the possibility that murderers too might have read *The Pale Horse*, although Christie herself, thorough as ever, went out of her way to express the hope that they had not. The population at large, meanwhile, remains happily ignorant of the effects of thallium. What else could explain the decision by the perfumer Jacques Evard to launch a men's fragrance called Thallium, a product whose implied promise includes baldness and impotence?

## *The Light of the Sun*

The search for the elements has always been an edgy business. It happens at the edges of recognized scientific disciplines and at the edges of respectable enquiry. New elements have been found as by-products of the alchemical quest for gold and the philosopher's stone. Discoveries have been claimed long before there was tangible evidence of pure new material, from the mere colour of a flame or when some inexplicable residue was left after a standard chemical analysis. More often than you would think, these finds have been shown later to be no more than fancies based on these brief, freakish observations and the vain ambition of the would-be discoverer. You could compile a parallel periodic table of a hundred elements that were named in hope and yet never seen. But the story of one element suggests that forgiveness may be more in order than condemnation for those investigators who found themselves caught in these thickets.

Since the spectroscope had revealed new elements in the flames of humble salt and tobacco ash, it was entirely to be expected that before long somebody should take the hot new tool of chemistry and turn it towards the sun. In 1868, the French astronomer Pierre Janssen travelled to the Bay of Bengal to observe the total solar eclipse that would give science its first opportunity to probe the solar atmosphere. Disembarking at Madras, he was greeted by the British governor of the province and invited to set up his observation station where he wished. He chose the cotton town of Guntur, which lay in the middle of the path of the eclipse and nestled between the sea and the mountains, where mist and cloud were unlikely. It rained for several days leading up to the eclipse, and Janssen began to fear that he might have lugged his gear halfway across the world for nothing. However, according to Janssen's account, on 18 August, 'the day of the eclipse, the sun shone at rising, although still in a bed of mist; he soon emerged from it, and at the moment when our telescopes gave us notice of the commencement of the eclipse, he shone out in all his brilliancy'. Then, as the darkness enveloped the waiting observers, Janssen recorded: 'Two spectra, composed of five or six very bright lines, red, yellow, green, blue, and violet', arising from two 'magnificent protuberances' in the corona on either side of the sun at the moment of total eclipse. To the eye, this light appeared not white like full sunlight, but like 'the flame of a forge fire'. The spectroscope, however, saw discrete lines of colour separated by regions of black, which made it a simple matter to compare them with the spectral lines produced by known elements that had been confirmed in laboratories. While the red and blue lines matched the light – seen also in the normal solar spectrum – emitted by hot atoms of hydrogen, the yellow line did not. Though close in colour, it did not correspond precisely to the characteristic yellow of sodium either. Janssen concluded that this line must be owing to the presence of an unknown element, though, perhaps foolishly, he was not bold enough to give it a name. A couple of months later, the British astronomer Norman Lockyer observed the sun through the autumnal Cambridge sky and, comparing his findings with those from a discharge tube of hydrogen (the principal solar gas), arrived independently at the very same conclusion. Thinking the element might be present only in the sun and not found on earth, Lockyer named it helium after *helios*, the Greek for sun.

With no hard evidence to support their audacious claims, both Janssen and Lockyer faced years of mockery during which irreverent scientists would quip to one another of this or that unknown concoction, 'that's helium'. Many spectroscopists doubted whether helium truly existed, and even Edward Frankland, the chemist who had assisted Lockyer in his experiments, continued to believe that some undiscovered emission of hydrogen was a more likely explanation of the yellow line. It was not until 1895 that Lockyer was finally vindicated, when William Ramsay was able to send him a discharge tube full of helium gas that he had gathered from the radioactive decay of a uranium mineral. Lockyer rejoiced: 'the glorious yellow effulgence of the capillary, while the current was passing, was a sight to see'.

In the meantime, before Ramsay could come to the two men's rescue, other astronomers had gaily begun reporting further discoveries of celestial elements that were beyond the reach of any confirmatory laboratory test. Coronium was claimed in 1869, and only in 1939 proved to be iron. Nebulium followed, but turned out to be an energized form of oxygen. It was Mendeleev's organization of the periodic table – and the gradual filling of its vacant spaces since – that finally put an end to these wild claims. There remain many unidentified spectral lines in the annals of astronomy, but the chances that any of them are owing to an undiscovered element rather than to uncatalogued electronic excitations in known substances are now zero.

*

Less reputable investigators, however, have been keen to exploit the air of mystery that often clings to newfound elements – or elements, as it might be said of helium, uncomfortably stalled on the threshold of discovery. To the lay observer, after all, the coded evidence of the spectroscope surely appeared hardly more credible than the ravings of a cabalist. In a scientific age when people were asked to believe in invisible X-rays that could see through solid matter and radioactivity that could magically cause one element to be transmuted into another, any novelty seemed possible. And if elements were to be found beyond the range of human perception by looking to the heavens, then was it not reasonable also to seek for them closer to home by more congenial extra-sensory means?

The case of occultum presents this other side of the balance sheet – an elemental find claimed not by learned men of science but by self-avowed mystics, yet reliant, just like Lockyer's claim to helium, on visual evidence produced by arcane means and only directly witnessed by a select few observers.

Occultum was the 'discovery' of Annie Besant and Charles Leadbeater. Besant was a leading light in the theosophist religious movement, a clairvoyant, a feminist activist and a leading political radical of the Victorian period. With Leadbeater, a former Anglican preacher, she wrote many books, among them one called *Occult Chemistry*, a fusion of these later interests with what she had learnt while studying chemistry as one of the first woman undergraduates at London University. This volume, first published in 1909 and later running to several editions, gave exhaustive and precise descriptions of the appearance of individual atoms of many of the elements as they appeared first to Leadbeater and then, under his tutelage, to Besant, viewed by the 'third eye' of clairvoyance. The atoms were illustrated by Curuppumullage Jinarajadasa, Leadbeater's young Singhalese companion, who attended the chemical séances along with his

white kitten. He did not see the atoms himself, but made beauti-
fully detailed drawings of them based on Leadbeater's and
Besant's descriptions. They looked uncannily like the spirogyra
and spicular marine organisms illustrated by the German biolo-
gist Ernst Haeckel, whose magnificent compendium *Kunstformen
der Natur* had been published not long before.

Leadbeater and Besant launched their eccentric atomic
project in 1895. Besant, remembering her student days, stated
the importance of observation above all, and made a show of
reporting neutrally what they claimed to see. They started
with an attempt to observe 'a molecule of gold' but apparently
found it 'far too elaborate a structure to be described'. Leadbeater had better luck with hydrogen, which he announced had
a countable number of minor atoms 'arranged on a definite
plan'. This simplest of elements 'was seen to consist of six small
bodies, contained in an egg-like form. It rotated with great
rapidity on its own axis, vibrating at the same time, and the
internal bodies performed similar gyrations.' It was found to

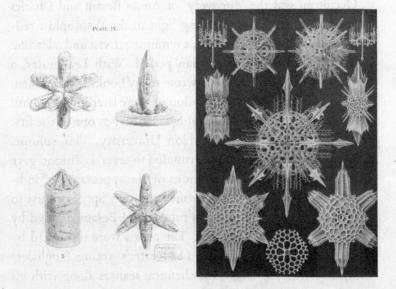

PLATE IV

weigh eighteen *anu*, a unit of measurement devised by the occultists, who named it after the word for the indivisible unit of matter in Jain metaphysics. Leadbeater and Besant observed elements more elaborate than hydrogen but less daunting than gold and 'weighed' them too. Nitrogen and oxygen were found to measure 261 anu and 290 anu respectively. The agreement between these numbers and the two elements' relative atomic weights, as determined by more conventional means, was quite remarkable.

That same year – the year, it should be remembered, when Ramsay confirmed the terrestrial existence of the elusive solar gas – they also observed an atom 'so light, and so simple in its composition, we thought that it might be helium'. Unable to get hold of a verified sample of helium, however, they admitted they were unable to confirm this attribution. In 1907, Leadbeater and Besant did finally obtain some helium gas and subjected it to the mysterious scrutiny of the 'third eye'. They pronounced themselves surprised that 'it proved to be quite different from the object before observed, so we dubbed the unrecognised object Occultum, until orthodox science shall find it and label it in proper fashion'.

Orthodox science never did find it, of course; occultum duly went the way of coronium and nebulium. Yet Besant and Leadbeater cannot simply be dismissed as cranks. They consorted with scientists. They observed and measured, and recorded their observations and measurements, with great thoroughness, just as scientists do. Furthermore, leading scientists were not unknown to dabble in alternative religion. William Crookes, the discoverer of thallium, was a fellow of the Theosophical Society himself and on occasion provided samples and advice to the occult chemists.

On the other hand, Besant's and Leadbeater's research does fail the first test of experimental science inasmuch as nobody has been able to replicate their results. Recently, Michael McBride, a chemist at Yale University, looked again at their data and subjected it to a statistical analysis. He found that the agreement between their figures for relative atomic weights of the elements and those accepted by science was not just close, it was too close to be true: any genuine experimental procedure would have produced a wider spread of data. McBride clears Besant and Leadbeater of fraud, however. He believes instead that a collective delusion led them to associate their 'observed' values with the established ones.

They plainly did not see individual atoms as they claimed, yet compared with so much else happening in chemistry and physics at the time, you could argue that this actually made their results appear more scientific rather than less so. (X-rays, also discovered in 1895, would eventually enable scientists to 'see' atoms.) The plausibility of Besant's and Leadbeater's claims is further enhanced by the detail of their accounts, their insistence on the pieties of science ('it is very desirable that our results should be tested by others who can use the same extension of physical sight'), and their irresistible illustrations; illustrations which look like strange sea creatures, yes, but also – and this is uncanny – very like the diagrams of the orbits of electrons around atoms and molecules that were much later devised as an aid to understanding the nature of chemical bonding. Though it was certainly not its tellers' intention, the story of occultum might almost be considered as a satire on the rhetoric of scientific presentation, gussied up as it is with its technical terms, lengthy exegeses and elaborate visualizations of what, in fact, cannot be seen.

There are moments when Besant's and Leadbeater's imagined system of the elements based on the recurrence of certain subatomic shapes comes across as plain crazy, as when they write, for example: 'Manganese offers us nothing new, being composed of "lithium spikes" and "nitrogen balloons."' Yet the great Crookes, admittedly careful in his praise, recommended that 'their work would be useful at least in suggesting to scientists the kind of elements they might still discover in the as yet unfinished periodic table'. In the event, their visions came closest to reality in atomic physics. Besant and Leadbeater believed that even the simplest atom, hydrogen, was composed of many subatomic particles, and that both the atoms and their constituent particles were constantly spinning and vibrating – all phenomena that would be observed by physics during the next few decades, the spin of the electron revealed, in fact, by examining the detail of the helium spectrum.

The intangibility of helium finally got to Lockyer. Not satisfied with Ramsay's gift, he sought to obtain his own sample of the element and in 1899 wrote off for promising source materials. In reply, the superintendent of wells and baths at Harrogate sent Lockyer some salts from his spa. The waters of such places were known by now to fizz not only with hydrogen sulphide and carbon dioxide but also with small amounts of the inert gases. Carefully collecting the gas released by the salts, Lockyer at last held in his hand the element he had detected more than thirty years before.

# Part Three: Craft

## *To the Cassiterides*

The Phoenicians sailed far and wide in search of tin. They probably obtained the metal first from sources in Crete and Turkey, then, ranging west, from Etruria in Italy and Tarshish in southern Spain, and east, from as far away as the Malay peninsula, where much tin is still smelted today. But their most fabled source was the islands known as the Cassiterides.

The Phoenicians flourished in the land that is now Syria and Lebanon for more than a millennium beginning around 1500 BCE, promoting trade and technological development but leaving few records of their doings. It is the Greek writer Herodotus who is largely responsible for the myth of the Cassiterides, the place to which the metal is for ever linked by the name of its ore, cassiterite. Though he personally doubted the islands' existence, he nevertheless wrote them into his *Histories* around 430 BCE, and so, true or not, into history:

Of the extreme tracts of Europe towards the west I cannot speak with any certainty; for I do not allow that there is any river, to which the barbarians give the name of Eridanus, emptying itself into the northern sea, whence (as the tale goes) amber is procured; nor do I know of any islands called the Cassiterides, whence the tin comes which we use. For in the first place the name Eridanus is manifestly not a barbarian word at all, but a Greek name, invented by some poet or other; and secondly, though I have taken vast pains, I have never been able to get an assurance from an eye-witness that there is any sea on the further side of

Europe. Nevertheless, tin and amber do certainly come to us from the ends of the earth.

Yet there is sea on the farther side of Europe, and the Cassiterides must exist, for tin was indeed brought to the Mediterranean from the west, the trade being run from the Phoenician port-state of Carthage. But where in the west? The mystery may be deliberate. Pliny the Elder in his *Natural History* writes that the metal came from 'Lusitania' and 'Gallaecia' and was also 'brought from the islands of the Atlantic sea in barks covered with hides', while the Greek geographer Strabo, writing 400 years after Herodotus, suggests that the Phoenicians may have deceived their enemies as to where these valuable resources lay, but hazards that these islands lay off the Iberian coast 'to the north of the port of the Artabrians'. But there are no such islands. Later scholars have interpreted Classical accounts as references to the north-western extremity of Spain itself, or Brittany, or the islands at the mouth of the Loire and the Charente in the Bay of Biscay. But these places have no tin. So far, so unreliable, and after all, as one modern metallurgical text tartly reminds us, 'how many historians of our day could tell us whence we derive our tin?'

Another Atlantic promontory *is* rich in tin, but then Cornwall is no island. Perhaps we are taking the third-hand reports of ships' lookouts too literally. For Mediterranean scribes, it would have been a superfluous act of imagination to give definite shape to any extensive land reported from voyages into the endless ocean that lay beyond the Straits of Gibraltar; how much more fabulous simply to conjure an island. And more plausible too, for who would believe it more likely that the Phoenician ships had simply doubled back on themselves to discover no more than the far side of a continent they already knew?

Tin has been exploited in Cornwall since at least 2000 BCE,

obtained from river-beds or by setting fires directly against the rock to melt it out, and thus was long established by the time Phoenician traders heard of it. Yet the idea that the Cassiterides, known to the ancient world so specifically as the *Tin* Islands, and 'ten in number' according to Strabo, were truly islands rather than parts of a larger mass of land cannot be so easily discarded. The logical assumption that they may be the Isles of Scilly seems to fall at the first hurdle – they possess very little tin. I ask Richard Herrington at the Natural History Museum in London what he makes of the competing theories. He favours the idea that the tin did come from Cornwall and that the Isles of Scilly served as a convenient trading centre. Here, inshore craft – Pliny's 'barks covered with hide' – might meet the large ships of the Phoenician traders who, sailing north past Cape Finisterre ('Artabria'), might just consider the Scilly islands as lying off the coast of Spain. This scenario at least reconciles the historians' descriptions with the mineralogical facts. The Phoenicians need never have seen the British mainland.

There is another dimension to the mystery of the Cassiterides – their name. The standard view is that the islands are named after the valuable ore found there, but some have wondered whether the boot is not on the other foot, and the ore takes its name from a pre-existing name of the islands – much as it is believed that the Latin word for copper, *cuprum*, may derive from Cyprus, the place which was the Mediterranean world's main source of this element. This seems rather unlikely – the Sanskrit word for tin, *kastira*, points to an Indic etymology based on Asian sources of the metal. But this ancient root does at least underline the claim of Cornwall to be among the very oldest known sources of tin.

I have a modern map which, although it does not claim Cornwall as the Cassiterides, does show it to be a land of tin. It is a 'metallogenic' map of the British Isles – it tells where the nation's treasure is buried. The land area is tinted in pastel colours to represent the major geological periods, and on top of this are scattered little coloured lozenges like a spilt pick'n'mix. The scatter is notably uneven. It divides the country sharply in two: the bland Mesozoic to the south and east, and the Celtic regions to the north and west, where the geology rushes backwards in time through the Carboniferous to the Cambrian period and beyond. The coloured shapes cluster in these latter regions, denoting the presence of elements such as strontium at Strontian in Argyllshire, Welsh gold and many others. The shapes are designed to give an idea of the extent of each deposit and even to show which way the strata run. The spine of Cornwall is festooned with orange rectangles, which signify the presence of tin, tungsten, copper, molybdenum and arsenic. The largest rectangles are at the very end of the Cornish peninsula (although there are none on the Isles of Scilly). I decide I must make my own voyage to the Cassiterides.

It is immediately obvious that I am in a land of more interesting geology as I cross into Cornwall. Everywhere is evidence of quarrying and mining, white scars in the hillsides left by china clay workings, pointed slag heaps, the occasional mineshaft or chimney. The oldest, and now most picturesque, tin mines are on the rocky north coast of the Land's End peninsula. The area is now designated as a UNESCO World Heritage Site, placing them, incredibly, on a par with Easter Island and the Pyramids. Strangely, the ruined stone buildings live up to this honour, their conical stone chimneys and the blocky verticality of their shaft houses producing their own austere geometry.

There are many of these constructions littering the rugged landscape, but the surface buildings are the least of it. Underground, as I learn from an intricate wire model the size of a large room at Geevor mine, lies a complicated grid of tunnels

and shafts, a veritable underground city, constructed to follow the tin lodes wherever they led, sometimes even out under the sea. A tour of Geevor gives me a proper sense of the tin miner's lot. On the surface are the sheds where the ore was broken up and graded, the huge sloping rooms of shaking rhomboidal tables where heavy ore was sifted from the light, and the Piranesian horror of the calciner where arsenic was roasted off. Finally, we are taken down Wheal Mexico, one of the oldest parts of the mine workings, whose hard granite walls still ooze lurid blue copper. When we come 'back to grass', I am struck in a conceited twenty-first-century way by the incongruity of the breathtaking scenery with the hell of work below ground.

Next door to Geevor, the Levant mine reminds me of my purpose. Are these the Cassiterides? Levant, the traditional name of the eastern Mediterranean shore, seems too obvious a clue. It makes me suspect a certain amount of romantic postrationalization. If a shaft can be named Mexico in the hope that it might yield riches the equivalent of that country's silver, then surely other names may mean as little. But I am assured that the name dates back well over 1,000 years, and arose from commercial links with a Mediterranean trading company. I learn that in addition to the Isles of Scilly, the island of St Michael's Mount, then known as Ictis, in the sheltered southern bay between Land's End and the Lizard, may also have been a point of loading British tin for export.

Cornish tin was exceptionally pure, and maintained its reputation throughout Europe for centuries. Most of the mines only closed in the mid twentieth century – the reason they look so fine is not that they have been restored, but simply that they haven't had time to fall down. A few mines, such as Geevor, pressed on until around 1990, by which time the international tin cartel had collapsed and the price of the metal fell below three dollars a pound, making further mining uneconomic. Recently, the price of tin has recovered, which has encouraged hopes that mining may restart. 'Cornish people want to see it back,' David Wright, Geevor's assayer-turned-tour guide, tells me. 'It caused a great deal of misery, but it's part of Cornish history.'

Primo Levi calls tin a 'friendly' metal. He lists high among its amicable qualities that it gives us bronze, 'the respectable material par excellence, notoriously perennial and well established'.

The raw material for bronze in antiquity was copper ore that, unknown to the metalworkers of the day, contained enough tin to make the alloy. In many places, bronze and copper must have been thought of as distinct metals. There was no quest for the elements and no incentive to try to separate bronze into ingredients since it was already the superior metal for so many purposes. In a few places, pure tin was smelted from its own ore, cassiterite, and, too soft for weapons and utensils, was formed into ornaments. Where tin and copper were obtained from separate ores, it was naturally not long before bronze was being made purposely by putting the two metals together. Once it was known that bronze could be made in this way rather than relying on ores that happened to contain the right proportions of copper and tin, the hunt was on for the miraculous metal which had the power to make copper both more useful and more beautiful.

Yet it is not only as the essential ingredient of bronze that tin has found its role. The metal has its own advantages. Unlike lead, it is shiny and bright. It is strong enough to make useful items, yet soft enough that these articles may be formed by simple hammering, demanding no great artisanal skill. Above all, it is easily smelted and cast, melting at 232 degrees Celsius, far lower than copper or silver.

I knew this from repeatedly melting and casting the same piece of tin in different shapes as a boy. But I am reminded of it when I go along to a workshop intended to reacquaint designers and academics who spend their days creating computer graphics or completing student assessments with the quiddity of real materials. Our tutor for this exercise in casting tin is Martin Conreen of Goldsmiths College in London. Conreen has the twinkle in his eye to make a good department-store Santa Claus, although his beard is ginger. Gleefully, he reaches into his sack and distributes his metallic hoard, a small, shining ingot of tin for each of us – and a cuttlefish bone. Cuttlebones, Conreen explains, have been used at least since Roman times as moulds for tin ornaments. We eye him doubtfully, but as soon as we start scraping away for ourselves we understand why. The porous bone is easily carved but also able to withstand the heat of the molten tin. Carefully, we melt our tin and pour it into the recesses we have carved in the bone. After a few moments cooling, it is possible to turn out the trinkets. Their weight and silvery lustre make them a delight to hold. The molten metal has faithfully followed every runnel I carved and has even picked up the fine honeycomb texture of the cuttlebone itself, adding a serendipitous layer of natural ornamentation. The satisfaction of making something so solid and pleasing tells in the silly grins on our faces.

Because tin is so readily recast, it has a special value in storytelling: it may be recast in different roles. Hans Christian Andersen's fairy tale of 'The Steadfast Tin Soldier' ends tragically with the soldier being consumed in a fire along with the paper ballerina he loves. Raking the ashes later, a maid finds that the soldier has melted into the shape of a heart. Tin is disposable yet also indestructible; the tin soldier is mortal, but his love endures. He is also cruelly subject to fate. The boy who features in the story has thrown him 'without rhyme or reason' into the fire. And there is a hint at the beginning that fate will be playing a part as the tin hero is already different from the other twenty-three soldiers in the box: he was made last when the metal was running out and has only one leg. This fatal thread in the tale, too, is drawn from the way the metal is handled: the die is cast at the beginning and recast at the end.

Its ease of working made tin the commonplace metal. Bronze was reserved for weapons, gold and silver for the Church and court. Ironware required the services of a blacksmith, but anybody might work a piece of tin into something useful. For the peasant, tin stood in for all of these metals for ornament and utensil alike and was made into plates, jugs and tankards, musical instruments, jewellery and toys.

Tin was also ideal for making prostheses which could be moulded and beaten to follow the intricate shapes of the body. The phrase 'tin ear', meaning tone-deaf, dates from the days when people could all too easily lose an extremity to the ravages of the pox or some grim accident. (Copper noses were not unknown either.) The Tin Woodman in *The Wonderful Wizard of Oz* is a woodchopper whose axe is bewitched such that it severs his limbs one after another and then finally lops off his head. Each time, he is given replacements of tin – though his chronically rust-prone joints suggest that metallurgy wasn't Frank Baum's strong suit.

Medieval craftsmen worked with tin as it came: smelted from the ore without further refining. Though Cornish tin was renowned for its purity, even this metal was frequently contaminated with lead, copper, antimony and arsenic, which affected its properties – often for better, sometimes for worse. Later bismuth was added in small amounts to convert the soft metal into a harder, more lustrous and sonorous alloy. Bismuth indeed was thought to be a mixture of lead and tin until proper chemical investigation in the eighteenth century showed that it was a distinct element. (Today, most of us encounter bismuth only when we have an upset stomach: it is the pepsin-zapping active ingredient of Pepto-Bismol.)

The alloying of tin was the secret of the pewter smith. Pewter today probably conjures up images of christening mugs of dubious taste or the engraved tankards of the regulars in unreconstructed pubs. Historically, it was an alloy mainly of tin with lead, which fell from favour when the poisonous nature of the latter was understood. Today pewter is entirely made from tin with a little antimony, bismuth and copper. The Worshipful Company of Pewterers is one of the oldest of London's guilds, with its origins in the fourteenth century, but it is making a determined effort to rehabilitate the metal with an annual competition for designers, who respond valiantly with necklaces and wine coolers and light fittings that do indeed succeed in slipping the shackles of history.

While pewter struggles to retain its foothold as an attractive material, tin has become a pejorative term for any cheap metal. Low-value coinage, usually based on copper, is 'tin'. Henry

Ford's Model 'T', the most basic of automobiles, made of steel, was the Tin Lizzie. Hastened by the growth of tin-plating during the nineteenth century, tin has stretched to become a looser metaphor for anything superficial or contemptible, for the marked-down and the jumped-up. Rudyard Kipling invented 'the little tin god' as an epithet for any petty despot in his *Departmental Ditties* of 1886. 'Tin-pot' continues to be used as an adjective, reserved almost exclusively for foreign dictators, to describe one who is reliant on the finery of his office to disguise his underlying corruption – British leaders were naturally never tin-pot themselves.

The metaphor is hardly fair to the metal, as the tin of the tinned cans which fed the British Empire was there to prevent corruption in the first place. On one occasion, as part of a student lecture at Guy's Hospital in London, a tin of meat left over from a naval expedition in 1826 was opened some twenty years later. The contents were found to look and smell good, so good in fact that they were swiftly consumed by some passing hospital staff.

Tinkers, too, are regarded with suspicion. The label is popularly supposed to apply to tinsmiths, but really refers to any itinerant mender of utensils. Somebody who tinkers is generally guilty of inept or uncertain handiwork. The dignity of the itinerant and of tin is reasserted, however, in Rose Tremain's recent novel *The Road Home*, which describes the experience of Lev, a migrant from Eastern Europe who travels to Britain for work. Tremain subtly opposes two elements, sodium and tin, in her narrative. Lev's bus crosses into Austria during the night and stops for petrol under a 'sodium sky', a recurrent image. Back in Poland, his grandmother supports his family by making jewellery out of tin. Sodium signifies the modern, technological sophistication, the urban West. Tin speaks of home in the rural East of simple crafts, a world Lev evokes so fondly that even his

Irish flat-sharer considers moving there. Tin, like Lev in the story, is compliant and cheap but nevertheless fundamentally honest and decent.

Tin is said to cry when a stick of it is bent or broken, which Primo Levi gives as one further reason for considering it a friendly element, even though he doesn't appear to believe it – 'never seen or heard (that I know) by human eye or ear'.

Levi's ignorance and incuriosity on this occasion is a puzzle. I have certainly heard this weeping, and I heard it again during Martin Conreen's materials master-class as I tortured my own piece of tin – a drawn-out cracking sound with a squealing overtone, like the opening of a door in a horror film, as the crystals of metal were wrenched apart. In truth, the phenomenon is not even unique to tin but can be produced by stressing any suitably brittle metal.

The sound world of tin is special nevertheless. Its very name rings: *tinnnnn*. And this is not by chance. As the metal most commonly made into domestic utensils, tin brought sonority into ordinary lives. The ringing of bells and gongs confined to Church and state rituals received a humble domestic echo in the sound of tin on tin in people's homes. The quality of the metal was measured by the purity of its ring. The onomatopoeia is widespread. The English word comes from the Old High German *zin* – it's still *Zinn* in German, and has similar names in other Nordic languages. The French word, even though it derives separately from the Latin *stannum*, is the almost homophonous *étain*. The word 'zinc', it is worth noting in passing, may also stem from *zin*, which may have something to do with the uncertain chemistry of a time before it was known that zinc was an element distinct from tin. Lead, incidentally, Shakespeare's 'dull lead', is named with equal onomatopoeic truth because it does *not* ring, the more so in old Scandinavian languages where it is called *lod*.

Zoe Laughlin, a materials scientist at King's College London, has made a study of the characteristic sounds made by different materials, going to the trouble of making identical tuning forks in glass and wood as well as various metals. She then noted the sound they made when struck, recording objective measurements of pitch and timbre, loudness and attenuation, as well as asking a panel of musicians for a more subjective assessment. She found that steel forks produced the brightest sound at the highest pitch. Copper and brass sounded deeper but almost as bright. For reasons that remain to be explored, the brightest tone of all came from a steel fork that had been plated with gold. Sadly, a fork cast in solder, which is mainly tin, failed to produce a tone, and soon exhibited signs of metal fatigue: whether it cried is not recorded.

Many of the words to do with ringing metals are generic; it

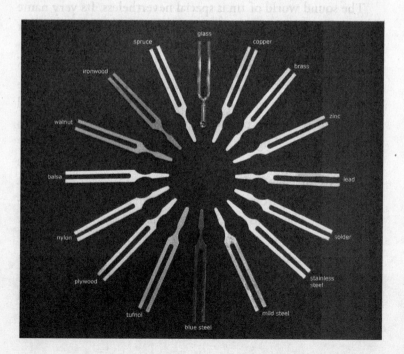

doesn't matter which metal is being beaten. Tinkers then are not exactly tinsmiths, but so called from the tinking sound made by their working with metal, be it tin or something else. In Spain tinkers are known as *quinquilleros* and French ironmonger's stores are *quincailleries*, the words amply expressive of the clanking sound of any metal merchandise. Norse and Germanic myth makes a strong connection between smithing and song, and we still speak of 'hammering out a tune', something Wagner depicts literally in Alberich's subterranean forge in *Das Rheingold*.

Other sense associations are specific to tin, notwithstanding Laughlin's disappointing result with the tuning fork. Some are physically connected with the material's properties, but others reach figuratively deeper into the aural world. Most majestically resonant of all, organ pipes are traditionally made of tin alloyed with lead, the proportions variable according to the tone that is desired. At the other end of the spectrum, tin whistles and tin drums need not be made of tin but certainly have the sound that we characterize as tinny. Tin Pan Alley in New York got its name from the noise made by the pianos of composers – tune-smiths – banging out popular melodies. Even tinnitus, the sensation of ringing in the ears, joins this clangorous family.

Let us leave this element to the sound of bells, or tintinnabu-lation. Bells can be made of any sonorous metal – my friend Andrea Sella even has one made of mercury waiting in a deep freeze at the chemistry department of University College Lon-don for the cold day when it will be rung.* But it has long been known that an alloy of copper and tin in the proportions of three or four to one produces the best tone. This special bronze is brittle and notoriously difficult to cast, and a number of old fables turn on the fortunes of the person who possesses the

*Since writing this, I have been informed that the bell will not now fulfil its destiny, having met an ignominious end when it melted during a power cut.

secret of bell-making. Many bells have cracked, among them the Liberty Bell, which, having safely crossed the Atlantic in 1752, split from lip to waist upon its first striking in Philadelphia. Big Ben, too, cracked shortly after it was installed in the newly built Houses of Parliament in 1859. The pompous bronze of statues may be respectable par excellence, as Levi suggests; bell metal, with its greater proportion of tin, rings out the happy imperfections of humanity.

## Dull Lead's Grey Truth

During the 1880s, Auguste Rodin, the most famous and controversial artist of his age, created what would prove to be his most popular work, *The Thinker*. It was intended as the central feature of a much larger composition, *The Gates of Hell*, which was to serve as a monumental portal for the new museum of decorative arts in Paris. The massive work, nearly seven metres high and seething with humanity, was never finished to the artist's satisfaction, but parts of it, including *The Thinker* (originally envisaged as a figure of Dante), were eventually finished separately on an even larger scale. The pose – hand propping chin, elbow rested on knee – may be overfamiliar now, but the sculpture still has the power to rise above parody. The figure leans forward impossibly far. The cantilever – which would have been still more dramatic viewed from below as you passed under the lintel of the museum doorway – is crucial to Rodin's achievement. This static lump of bronze is animated even by Rodin's usual standard, producing not an outward appearance of movement as sculptors often sought to do, but a projection of internal activity. It urgently wants us to know *something*, to know in fact the very power of thought. Recent X-ray studies have shown that the sculpture is only able to do this to such an extraordinary degree because it conceals within its base a massive counterweight made of lead.

Lead is the reification of gravity, both physical and intellectual, and is the chemical element most closely associated with death itself. When we speak of a leaden sky, it is not only the colour we mean: the gravitational impossibility of the image presages worse than rain – the doom of a world turned upside down. Lead sarcophagi are traditionally used to preserve the bodies of popes and kings to ensure that the soul does not escape. The heart of the king of Scots Robert the Bruce rests in a lead casket at Melrose Abbey, as does the lanky body of his foe, the English King Edward I, at Westminster. The 'Hammer of the Scots' instructed that the casket was to be exchanged for regal gold only upon the final defeat of Scotland; the lead casket remains to this day.

Lead does not corrode, and so preserves what it contains, because it forms a surface layer which blocks further chemical attack. It is this thin layer – the same substance as artists' lead white – which ultimately preserves the roofs of many of the cathedrals and churches of Europe as well as the bodies of their prelates. This compound also robs the metal of what little lustre it has when fresh cut, leaving it an elephant-grey that hardly reflects sunlight. This, too, seems to render lead more suitable than others for rituals of death and burial.

Lead's weighty relationship with gravity and its connotations of the ultimate collapse – into the tomb – are but the most extreme of its various associations with fate and falling. When we agree to leave a matter to chance, we let the chips 'fall where they may', governed not by us but solely by the laws of physics. One of the secondary meanings of the German noun *Fall* is simply 'event', something that happens or befalls. And a fall gains emphasis if what falls does so heavily. A heavy fall is decisive. For this reason the Romans made dice out of lead.

In parts of central Europe where lead ores are abundant, the custom has grown of predicting the future by pouring small quantities of the molten metal into water. The metal solidifies naturally in extravagant shapes, and it is from these that the pourer's fortune is deduced. Germans perform this ceremony of *Bleigiessen* (lead-pouring) on New Year's Eve. If the solidified lead resembles a flower, then you will enjoy new friendship in the coming year. The shape of a pig foretokens prosperity, a ship a long voyage, and so on. In Hungary, the ceremony takes place on Luca's Day (13 December), when lovers pour lead to divine the qualities of their intended partners. The traditions still thrive and, surprisingly perhaps, you can easily buy children's kits containing real lead for melting and pouring at home.

Certainly, it's a procedure that doesn't seem to require an expert on hand to interpret the results. I decide to improvise for the benefit of my family using some lead salvaged from an old leaded window. Heated in a ladle by the flame of a portable bunsen burner, the crumpled metal quietly collapses until it quivers beneath a layer of white and yellow oxide and is ready for pouring. Can you really tell fortunes, wonders my nine-year-old son as we watch. He goes first. I pour about a dessertspoonful of molten lead into a bucket of water and he retrieves one of the bigger pieces. It is pear-shaped, and we are at a loss for any career it might portend. Then he turns it the other way up and declares that it looks like a balloon. Perhaps he will travel the world by air. My wife is next. With more practised pouring, more elaborate shapes are formed. She picks out an elongated dribble that miraculously does indeed resemble a flower on a stem. A new friendship in the coming year seems a safe enough bet. Finally it's my turn. I pour the lead again, and extract from the water some attenuated lumps that fail to inspire.

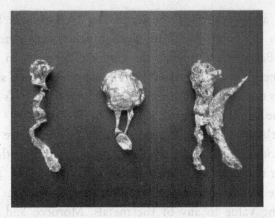

But one more sculptural fragment offers greater scope for the imagination, conceivably suggesting a human figure. The resemblance is marred by a splat of lead fused diagonally across the middle of the torso. Perhaps a musical instrument. Should I take up the lute?

Shakespeare recodes lead's predictive potential in *The Merchant of Venice*. In order to win the beautiful heiress Portia, her suitors must make an elemental choice between caskets of gold, silver and lead. Each casket bears an inscription. The inscriptions on the precious-metal caskets make promises that seem deliverable in some material form even if they are enigmatically stated. The golden casket is inscribed: 'Who chooseth me shall gain what many men desire', the silver: 'Who chooseth me shall get as much as he deserves'. The inscription on the lead casket recognizes only the uncertain world: 'Who chooseth me must give and hazard all he hath'. To choose lead now is acknowledgement that fortunes may not be told at all.

The first two suitors to appear in the play, whom Portia calls 'deliberate fools', are the princes of Morocco and Arragon. They judge that they cannot afford life's gamble; they prefer what is clearly a trade, even if the terms aren't clear. Vain Morocco chooses the gold, calculating Arragon the silver. The worthy suitor Bassanio rationalizes the choice differently. In a 'world still deceiv'd with ornament' he rejects both gold and silver and chooses the lead casket, opening it to find 'Fair Portia's counterfeit', the sign that he has won her.

All three suitors have been guided in their choice by their perceptions of the respective metals' worth. During the course of their deliberations, the men have called lead 'dull', 'base' and 'meagre' in turn, although Portia has been scrupulous in never assigning value to any of the metals. Morocco and Arragon allow the accompanying riddles to compound their confusion, but Bassanio, so far as we can tell, does not read the messages at all. His choice is corporeal. Morocco and Arragon are offended by the ordinariness of lead but Bassanio is undeterred.

Fortunes may not be for telling, but one thing in life can be foretold all too well. What Bassanio has won he knows too that he will ultimately lose. His correct choice indicates an acceptance of mortality – both his own and Portia's too. The lead of the casket has already spelt this out to all comers, as Morocco makes plain when he piously announces he cannot bear to 'think so base a thought' that Portia's portrait should be sheathed in lead in prevision of her death. Bassanio's paradox is that, while he desires the beautiful Portia, he can nevertheless confess of the leaden casket: 'Thy plainness moves me more than eloquence'. Lead's dull truth is that beauty fades. Time corrodes our bodies, our skin acquires its own oxide coating, but the soul may be kept pure within. The choice of the lead casket embraces this inevitability, showing that Bassanio will be a steadfast husband unto death. 'In this way,' wrote Freud in an essay on mythic three-way choices, 'man overcomes death, which he has recognized intellectually.'

'You that choose not by the view,' runs the scroll Bassanio

finds inside the lead casket, 'Chance as fair and choose as true! /
Since this fortune falls to you, / Be content and seek no new.' It
is a final reminder of the gravity of the decision he has made.

The fortune of all humankind is told in lead. The element's
traditional applications – many of them now performed by substitutes for health reasons – echo the ambiguous role it plays in
myth. Two of its oldest uses show how lead encompasses the
full range of human creativity and destructiveness: soldiers'
shot and printers' type. Balls of lead were used as slingshot pro-
jectiles in ancient times, but it was not until the discovery in the
fourteenth century that gunpowder, then new to Europe, could
be made to project a ball from a tube that the cannon became a
weapon of war. This crude device was gradually refined into a
broad range of firearms for which an equally diverse battery of
lead shot and bullets was required. Laboriously cast at first, lead
shot soon came to be made, as fortunes are told, with the assistance of gravity in purpose-designed towers. But unlike in
*Bleigiessen*, the element of chance is here carefully excluded.
Molten lead is poured from a height to form droplets of a certain size, which cool as they fall before being quenched in a
trough of water. I make my way to Crane Park on the western
outskirts of London, where one of these buildings still stands.
This tapered round tower was built in 1823 for the Hounslow
Gunpowder Mills. Today, it has been restored and lies pictur-
esquely situated on the edge of woodland with parakeets
swooping noisily in and out of the cupola at the top. A shallow
river rushing by provided the essential water. Standing on one
of the six circular galleries that gird its bare brick interior, it is
easy to imagine the hot lead falling down the middle and sputtering into the water below. A long fall – the tallest shot towers
were more than twenty storeys high – ensures that each piece is
close to spherical by the time it hits the water, but even then

further work is necessary to sort and grade the shot. Gravity is put to use here, too, as the spherules are set off down an inclined plane towards a kind of jump. Those that roll well are able to jump the barrier, while the oversized and misshapen sluggards fall short and are collected for remelting. (The luck factor returns when the shot is fired: though billions of lead ballistics have been manufactured and fired in anger, they have claimed mere millions of lives. This low strike ratio is declining still further, according to the experts, for the simple reason that technological advances in firearm design make it so easy to pull a trigger prematurely.)

One of the several distinct innovations of Johannes Gutenberg that leads us to claim him as the father of printing was his adoption of lead for type. Gutenberg had some training as a goldsmith and was a skilled metallurgist by the time, around 1440 while living in Strasbourg, he turned his mind to the problem of printing. He saw that the presses used locally for wine-making might be

adapted to press lettering on to paper, but for the lettering to be changeable so that different texts might be printed, he would need a material with special properties. It would have to be highly mouldable to take the intricate shape of each letterform, but also sufficiently durable to bear repeated impact on the paper. For fully movable type, each small piece comprising a single letter would furthermore have to remain loose so that, once released from the press, it could be rearranged to set new text. Gutenberg's answer – also reached independently in Korea around the same time – was to use lead alloyed with tin and a little antimony. This made the metal flow better while molten and yet form a harder letter when solid. This lead alloy proved ideal compared to bronze, which was harder to work with, or traditional materials such as wood blocks and clay, which were less durable. This 'type metal' dominated printing until the mid twentieth century, greatly accelerating the spread of knowledge and expanding the role of literature.

The profound and contradictory meanings of lead – fortune and fate, creativity and destruction, humour and seriousness, love and death – have led a number of contemporary artists to employ it in their work. Not many are drawn to such an unfashionable and humble material, perhaps, but the few who have been are among the most reputed. The British sculptor Antony Gormley and the German artist Anselm Kiefer, for example, use lead in ways that exploit contrasting aspects of its nature.

Kiefer works with an unusual range of basic, one might say primal, media including ash, chalk, straw and fingernails. Lead, which is regarded in alchemical and Cabalistic thought as primordial matter, has been important for Kiefer for more than thirty years, chosen for practical reasons of workability – it is one of the most malleable of metals – but also, more importantly, for its multiple cultural echoes. It is, he says, 'a material for ideas'.

In 1989, as East and West Germans began to chip away at the Berlin Wall, Kiefer was finishing a major work modelled on a modern bomber aircraft. Kiefer's plane is made not of aluminium, the lightest practical metal, but of lead, the heaviest. Its patchwork lead sheets are bent and folded into shape, and finished off with a crude parody of the bright rivets that we depend upon to carry us safely aloft. I see the work at the Louisiana Museum in Denmark, a place of harmony between land and sea, architecture and art, where it delivers a violent jolt like an injured bird seen on a country walk. In one sense, it is an hilarious proposition, a plane that could never fly. Like the lead axes made by the Romans it would be useless as a weapon of war. And, like the miniature lead boats that have been found on the Greek island of Naxos dating back to the Cycladic period 5,000 years ago, it is going nowhere. It promises flights of fancy yet remains heavily earthbound. Even its long wings and fuselage seem to slump, the spindly undercarriage barely able to resist the inexorable pull of gravity. The work is called *Jason*. In the

Greek myth, Jason and the Argonauts, whom he recruits to sail with him in search of the Golden Fleece, build a ship, the *Argo*, but find that it is too heavy to launch. It requires the magical intervention of Orpheus, who has joined the crew, before their voyage can begin.

Kiefer is interested in the fact that lead is mutable not only in physical ways; like us, it also seems to change its character. Many metals suffer from a phenomenon known as creep whereby they gradually deform under an applied stress. Lead is so dense and soft that it creeps under gravity alone, and Kiefer has exploited this property in works where ripples of lead pile up like waves on a beach at the bottom of the picture. Of the seven metals known in the ancient world, lead was regarded as the 'base' from which all the others were made in nature, and was the obvious starting point for alchemists striving to make gold. Kiefer believes the white-and-yellow crust that forms on the surface of molten lead is indicative of its 'potential to achieve a higher state of gold'. The element thus embodies hope, and Kiefer's works that employ it are intended to be expressive of hope for human-ity with its potential to change for the better. But to an artist born in 1945, the year in which the atom bomb was dropped, lead is also linked to a darker kind of mutability. Lead is the ultimate product of many radioactive decay chains, including those of the key atomic bomb ingredients, uranium and pluto-nium. In the old alchemy, lead speaks of potential for the betterment of humanity, but in the new it foreshadows its vio-lent destruction.

Antony Gormley's view of lead is informed by more familiar procedures. His 1986 work *Heart* is an irregular lead polyhedron. It alludes to the custom of preserving body organs in lead, and, coincidentally or not, also references the work of the German artist, for the same truncated cube recurs in a long-running series of Kiefer's works called *Melanchòlia,* inspired in turn by

Albrecht Dürer's engraving *Melencolia I*. The use of lead is apt
here as the alchemists equated the metal with Saturn, who was
the Roman god of melancholy.

Gormley's studio is a grand affair, walled and gated like an
embassy compound in a war zone. Inside, metal mesh human
figures hang by chains from the high ceiling. Light floods the
vast white space. I ask the artist about his materials. 'I like clay
because it is earth. I like iron in its original form as pig iron,' he
says. 'I am distrustful of bronze.' Whereas the alloy bronze is
charged with human artifice even before the sculptor sees it, the
clay of the earth and iron are elemental in one system or another.
Lead is equally basic. 'It is important to me that it's on the periodic table. I like the fact that it bridges the alchemical and
nuclear worlds.' Unlike Kiefer, Gormley coats the lead he uses
in order to prevent oxidation, which lends it a faint redemptive
gleam. In a work called *Natural Selection* (1981), familiar objects
– a banana, a light bulb, a gun – are sheathed in this anointed
metal. Human and other large forms are similarly treated in
other works, notably a series entitled *A Case for an Angel*, in
which each sculpture in the series represents a human body with
vast outspread wings, leaden forerunners of his steel *Angel of the
North* of 1998. These 'body cases' are hollow – the artist lists air
as one of his media in order that we understand this – so they
lack the heavy suspense of Kiefer's lead pieces. For Gormley, it
is the sarcophagal impenetrability of the lead that counts. We
are sealed out; air – and perhaps something more spiritual – is
sealed in.

Kiefer, on the other hand, prizes lead for its honesty. It
presents the unvarnished truth with all the ambiguous consequences that flow from it. 'It is, of course, a symbolic material,'
he says, 'but also the color is very important. You cannot say
that it is light or dark. It is a color or non-color that I identify
with. I don't believe in absolutes. The truth is always gray.'

*Jason*, the lead plane with its macabre cargo of human teeth and snakeskin, is one of several aircraft Kiefer has made that he calls his 'angels of history' in reference to the ideas of the philospher Walter Benjamin. Benjamin's 'angel' is a backward-gazing witness who sees history, not as we do as a sequence of passing events, but as an ever-accumulating pile of disasters, and who, despite wishing to, cannot go back and undo the damage because of the irresistible wind of progress blowing in his face. Kiefer worked on the sculpture as the cold war was coming to an end, a time when our safety was underwritten by such aircraft. The wind of technological progress had brought us to the point where our creative and destructive wills had converged in the supreme achievement of a high-tech machine for mass slaughter, and that same wind would now carry us on into the future with all its unknowable choices. Like so many of the leaden artefacts of the past then, *Jason* is a votive offering, one that expresses not only the bright hope that we will survive but also the dark fear that we will not.

## Our Perfect Reflection

In Richard Strauss's shimmering Mozartian opera *Der Rosenkavalier* (1910), the plot turns upon the moment when the amorous but essentially innocent Octavian gives the newly ennobled merchant's daughter Sophie a silver rose. An object of complex symbolism in an opera of symbols, the rose is meant as a customary token of nuptial engagement between Sophie and the boorish Baron Ochs. The seventeen-year-old Octavian has been persuaded by his worldly lover, the Marschallin, to act as the baron's emissary, the rose-bearer of the opera's title. Needless to add, Sophie is disgusted by Ochs but smitten by the handsome Octavian, who appears before her dressed for good measure in silver brocade. The drama proceeds, with the usual operatic confusions, to the young lovers' inevitable duet.

*The Great Gatsby*, F. Scott Fitzgerald's portrait of rich America in the jazz age, drips with gold but also with silver. The metal is present in images of the moon and stars and their reflections and in the opulent clothing worn by the fast-money millionaire Gatsby. It is both the sign of financial wealth and an indicator of its mineral provenance, for Gatsby's mentor Cody, we are told, is 'a product of the Nevada silver fields'. But silver is used especially to characterize the lively Daisy Buchanan, with whom Gatsby had years ago fallen in love – 'the first "nice" girl he had ever known' – before she married another man. Daisy is compared to a silver idol when they meet again, while the young Gatsby, not yet rich, was drawn to her in the first place both for her wealth and for her corrupted innocence, finding her 'gleaming like silver, safe and proud above the hot struggles of the poor'.

It is the same in England. In *The Forsyte Saga*, too, silver entangles wealth, class and the feminine. Soames Forsyte, 'the man of property' who gives the first in John Galsworthy's sequence of novels its title, collects and displays 'little silver boxes', which he holds equal among his possessions with his wife. 'Could a man own anything prettier than this dining-table . . . and quaint silver furnishing; could a man own anything prettier than the woman who sat at it?'

Silver has a deep cultural link with the feminine and with the moon, implicitly opposed to gold, which is equated with the sun and represents the male principle. This belief may not be quite universal but it is shared very widely by ancient cultures from Greece to the pre-Columbian Americas. The white lustre of the metal that explains these associations also carries with it more precise meanings to do with purity and virginity, and by extension with virtue, innocence, hope, patience and the passage of time.

For Baron Ochs, the silver rose is just an empty chivalric gesture (one with no basis in authentic custom, incidentally, but invented for the opera by Strauss's librettist, Hugo von Hofmannstal). In the hands of Octavian though, it becomes a potent symbol in which many of these meanings are simultaneously and confusingly present. The feminine aspect is especially highly charged as the role of Octavian, who must also appear at one point in the opera disguised as a maidservant, is sung by a woman.

These silver objects continue a thread that runs from the silver bow carried by Artemis, the Greek goddess of the moon and virginity and protectress of women, to William Blake, for whom there were 'girls of mild silver, or of furious gold'. But the element seems especially at home in the early twentieth century, during the years known as the Belle Epoque. By this time, households of even quite modest pretension could afford silver tableware thanks to the expansion of mining in North and South America, or if they could not, then there was at least silver plate. It was said of these new mines, as it was said of Mediterranean deposits during the Classical period, that the molten metal would run free from the soil at the source when forest fires broke out. Argentina – the only country named after a chemical element – was briefly the tenth wealthiest of the world's nations based on this resource.

Silver no longer possesses the social cachet that it did a century ago and its commodity price has plummeted. But, surprisingly perhaps, it has lost none of its symbolic value. The Silver Ring Thing, for example, is a movement begun in 1996 in the United States to promote chastity among Christian teenagers, although, acknowledging the reality of the situation, the 'para-church youth ministry' behind it all has made the strategic move, useful for recruitment no doubt, but unlucky for the symbolism, to admit not only the chaste but also the regretful, who are encouraged to 'embrace a second virginity'.

Silver also remains a familiar qualifier of branded consumer goods, where it is generally understood to convey a sense of pureness or even a cleansing property. The British Sugar Corporation produces a granulated sugar called Silver Spoon that betrays its customers' dyed-in-the-wool consciousness of social class as well as playing with ideas of refining and refinement. Silver labels products ranging from light beers and mineral waters to cosmetics, especially when these are targeted at younger women. It was entirely in keeping that Revlon should mark the twenty-fifth anniversary of its girlish perfume Charlie by rebranding it Charlie Silver, for example.

Perhaps because of its abundance of associations, and the fact that so many of them are tied up with the youthful vicissitudes of getting laid, silver is, according to a curious piece of research by Santiago Alvarez, a professor of chemistry at the University of Barcelona, the chemical element most cited in song. One of these songs, Don McLean's famous paean to van Gogh, 'Vincent', even manages an echo of *Der Rosenkavalier* with an image of a silver-thorned rose lying in the virgin snow.

★

Silver was the brightest and whitest of the elements known in antiquity. Its Latin name *argentum* is derived from the Sanskrit *arjuna*, which means white. This is no great claim from a time when so few metals were known. Gold and copper are coloured, which left only lead, tin and iron, which are all greyer, and mercury, which, though liquid and therefore often not regarded as a true metal, is nevertheless comparable in colour and so earns the name quicksilver. What is more remarkable, and helps to explain this element's enduring symbolism, is that silver is still one of the brightest and whitest elements in the modern periodic table, which contains more than eighty metals.

A polished silver surface has evenly high reflectivity of nearly 100 per cent across the full range of the visible spectrum of colour. It is the preferred coating for the mirrors in reflecting telescopes for this reason. (Aluminium, in comparison, reflects only about ninety per cent of light across the spectrum.) The reflectivity of silver dips slightly to ninety-five per cent in the violet, and this small diminishment of reflected violet light is what gives the metal its characteristic warm tinge of yellow. Silver, then, deserves its status as the pre-eminent gleaming white metal, and this quality alone could perhaps be held to account for its symbolic importance. But there is a further reason that explains why this element has retained and even consolidated its powerful significance through the ages of tinplate, stainless steel and chrome.

More than any other metal, silver signifies purity and especially virginity not simply because of its white lustre, but because of that lustre's almost human propensity to lapse into tarnished blackness.

Gold does not tarnish, which is why it is associated primarily with immortality. The alchemical symbol for gold is the never-ending line that is a circle, which represents not only the sun but also perfection. Silver's is a half-circle – an icon of the moon, but a symbol too of incompleteness or imperfection. Silver was regarded as incomplete simply because it wasn't (yet) gold. The alchemists reasoned that it wanted only for greater yellowness, which they sought to transfer from yellow materials as varied as copper, saffron, egg yolk and urine. The imperfection lay in its evident mortality, the tendency for a pure body of silver to become corroded over time and to end in a black death.

Unlike many metals, silver is not readily oxidized. But the sulphide coating that forms whenever a polished silver surface is exposed to sulphur in the air – which happens anywhere that candles or fires are burnt – is not brown like the oxides of iron and copper but a fine, deep black. A good layer of tarnish renders the surface of a silver object as black and matt – that is to say, as unreflective of light – as it once was shiny and white.

Silversmiths have traditionally sought to accentuate those qualities of the metal that enhance its connection with purity and femininity, favouring bright, smooth surfaces and fluid, voluptuous shapes. They were encouraged in their work by the metal's low melting point and high malleability, which make it easy to cast and cold-forge. Silver vessels intended for washing or drinking frequently depict water in their relief work and are decorated with such things as dolphins and mermaids. One especially extravagant eighteenth-century English ewer and basin in the Victoria and Albert Museum adopts the four Aristotelian elements as its theme and uses their contrasting properties to create a masterpiece of this sort of workmanship in which silver flames lick and silver rivulets flow with astonishing suppleness.

Even in more egalitarian times, silver remains a metal 'for items of luxury and decoration', in the words of one history of the material, 'best suited not to the monotony of a machine-induced finish, but to the caressing touch of the hand'. The mass-manufacture of silverware has now declined, and there has been a revival of craft interest in the metal. Craftspeople today, however, are as likely to work the metal against expectation as to stay within the boundaries of tradition. Silver is an especially tempting material for polemical or satirical treatment because it has for so long been identified with the upper classes. In 2008, I happened across an exhibition at London's Contemporary Applied Arts gallery called 'Tea's Up', a riotous display of handmade tableware that ripped apart the complacent niceties of the posh English tea party. China was broken and wrongly reassembled, silver spoons were reduced to crumbling wisps like archaeological fragments, cups and saucers were rendered as useless wire-frame outlines. One set of pieces was titled with ironical chutzpah after battle cries from the class war – ''Oi Polloi', 'Queenie', and so on. Others were named with memorable vulgarity after states induced by stronger drink than tea. A wobbly-legged silver jug called 'Trollied' – a colloquialism for being drunk – sticks in the mind. The author of these works, David Clarke, clearly struggles with the hypocritical virtue that hovers around silver. 'It's what I react to,' he tells me. 'At times, I get totally irritated by its almost religious associations. I respond in a devilish manner to corrupt the purity.' 'Trollied' turns out to be a relatively mild exercise. In other works, Clarke bakes silver with brine or mixes it with lead, which eats into it like a cancer. The resulting work is chemically alive, changing in response to the atmosphere. In summer, the salt causes copper from the solder to bloom green, while in winter the piece reverts to grey. 'It sets up a dilemma. What do you do: save the silver or enjoy the moment? Silversmithing is such an entrenched tradition. It is ripe to feast from. It is important for the future of silver that it has this chance. The discipline dies if it stays self-congratulatory.'

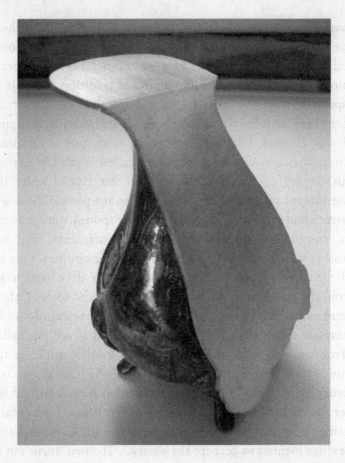

This project of subversion demands an exploration of silver's black 'other', and Clarke duly plans to turn his attention to tarnish – 'not the pure side of silver, but the dirty side!' Meanwhile, the artist Cornelia Parker has gone to the extreme of making tarnish alone the essence of the work. In a series called *Stolen Thunder*, she has rubbed the dirt film from various silver and other metal objects on to handkerchiefs. It is not beautiful art – they are just dirty handkerchiefs. But they are made more arresting with the information that the absent objects belonged to well-known figures – Samuel Colt's soup tureen, Charles Dickens's knife, Horatio Nelson's candlestick, Guy Fawkes's lantern. In some complex way, the tarnish seems to represent the price paid for the sparkle of celebrity. The easy chemical change of the metal to black tarnish and the effortful physical transformation back to shining metal by ritual polishing has written into it a narrative of death and resurrection, corruption and redemption. The handkerchiefs are evidence that Parker has spent time restoring some of the lustre to famous and infamous careers; and the viewer is invited to ponder the morality of that act. 'Silver for me is ten times more fascinating than gold because it has this duality about it and all the gradations between the two,' the artist tells me. 'You have to polish it to keep it shiny, and yet you're losing it, taking off a layer at a time. There is a taint about it, an original sin.'

It is not only its tendency to blacken that sullies silver's reputation, it is also that it passes through so many hands in the form of money. This usage of the metal deepens its ambivalence in

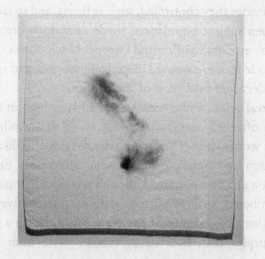

culture, as Shakespeare was aware. It is paradoxically the relative abundance of silver which has enabled it to fulfil this function. Gold, the obvious token of wealth, is simply too scarce. As coinage spread, it quickly became clear that there would never be enough gold to supply the demand for currency. Silver was rare enough to be valuable, but common enough to be a practical material for minting, and so this metal slipped into its now familiar role as the symbol of tradable value.

Emperors may lust for gold, but empires rise and fall in proportion to their access to silver. It was ironically the silver mines of Laurion at Cape Sounion that sustained Athens in its golden age. Later, a combination of slave revolts in the mines and expensive military campaigns against Persia meant that, in order to keep the economy going, the silver had to be stripped even from the Victory statues on the Acropolis. Finally, in 406 BCE, copper coinage was introduced.

The Romans, too, used silver for coin. They never really counted mining among their technological accomplishments, but they knew well enough how to exploit established mines in the territories they controlled, such as Iberia, and to take advantage when subject populations made new discoveries, as they did in the mountains of central Europe. Much of this newfound silver made its way eastward in exchange for silk and spices during the decadent final years of the empire.

The real price of silver reached an all-time high in Europe in the late fifteenth century, and this made the search for new reserves worthwhile. The Spanish discoveries of gold and silver in Mexico and South America soon afterwards funded the expansion of a new empire. Although it is the fabulous gold that is remembered, Spain imported six times as much silver in monetary value. The bounty of the New World led to a period of silver surplus that, boosted by further silver finds in North

America during the nineteenth century, continues today, with the result that silver now is worth less than one-hundredth of what it was at its peak in 1477.

Gold and silver are fairly interchangeable in the Christian liturgy. Goldsmiths habitually worked with both metals, silver was frequently gilded or alloyed with copper to make it look like gold, and gold and silver were used together to produce more decorative designs. All this helped to blur any distinction between the two metals. And in the yellow candlelight of a church interior, gold and silver start to look much the same – equally resplendent and generically precious.

More significant than the material of objects such as the chalices and patens used during Holy Communion and even the bishop's crozier was the style of their design and their degree of decoration. These could reveal a religious denomination at a glance. During the medieval period, goldsmiths vied to demonstrate their skill with pieces ever more elaborately crusted with ornament. But during the Reformation these fancy objects were seen as unacceptable 'popish plate' and melted down to be refashioned on plainer lines. Silver was now judged more seemly than gold, and it was finished without decoration, the gleam of a smooth expanse of the polished metal offering glory enough to God. As part of the same changes in liturgical practice, the congregation began to share in the Communion, which previously had been enacted by the priest alone. In the pure reflecting surfaces of the plainer silverware, worshippers might find themselves confronted at the height of the ceremony with that rare sight, in the days before mirrors were widely known, of the image of their own face framed in virtuous silver. And by drinking from silver, communicants may have received more than spiritual beneficence: chemical archaeologists have recently begun to recognize that the small amount of silver reacting with the organic ingredients of wine may have given it antiseptic properties, rather like the bacteria-battling silver nanoparticles that feature in today's refrigerators.

Although the Romans had discovered how to deposit silver on to glass in such a way as to produce a reflective surface, and the secret had been rediscovered in the Middle Ages, it was a skilled job to produce a surface large enough to check one's appearance, and mirrors remained luxuries beyond the reach of all but the nobility until well into the eighteenth century. Shakespeare's deposed King Richard II calls for a mirror that he may see himself 'bankrupt of his majesty'. He looks, and then dashes the glass to the ground: 'A brittle glory shineth in this face; / As brittle as the glory is the face'. When the Prince of Arragon opens the silver casket, he is dismayed to find not the likeness of Portia he is seeking but the 'portrait of a blinking idiot' – in short, he too sees himself in a mirror. He is the idiot for having chosen wrongly, finding only silver contained within silver, a looking-glass within the casket.

These two ancient qualities of silver – its propensity to tarnish from white to black, and the ability of its polished surface to reflect light so perfectly that one can see in it one's own face – come to a surprising convergence in the modern world. For like the mirror image, the photograph is an optical record captured in silver. From the beginning, pioneers of photography used light-sensitive salts of silver as their means to create black-and-white images. Yet strangely, nothing seems to have been written about the symbolic importance of silver, long established and widely agreed after all, in this major contemporary role. How does the choice of silver, the elemental embodiment of purity, virtue and the feminine, add meaning to the photograph? How do its values relate to the values of the camera's eye, its truthfulness and all-seeingness? Does the photograph, like the regal mirror, bring a necessary message of disillusion? Or does it have the power to purify the sitter? Certainly, from the very beginning, photography was pursued with each of these motives, as a means of documenting reality, and as a means of presenting an ideal. Yet when it comes to silver – the bridge between these two technologies of (human) image-making – the great commentators on photography such as Susan Sontag and Roland Barthes are silent. What sport they might have had with the chemical semiotics of the photographic process. For here, pure silver appears unexpectedly as the black knight, not the white. Photographic image-making depends on the chemical transformation of silver salts into silver metal by the action of light, and this time it is the pure silver, released first as single atoms and then as tiny clusters, that appears black.

It was in 1614 that one Angelo Sala, a physician from Vicenza, first recorded the natural darkening of nitrate of silver when exposed to sunlight. A century later, silver salts were being used to dye feathers and furs permanently black, and in 1727 Johann Heinrich Schulze from Magdeburg made photographic images of words by placing paper stencils over the surface of a bottle containing a mixture of chalk and aqua regia contaminated with silver. Despite this demonstration, and despite painters' widespread use of the camera obscura for the accurate rendition of landscapes, and despite even a detailed prevision of photography in Charles-François de la Roche's 1760 novel *Giphantie*, it seems that nobody thought to bring these optical and chemical processes together and record an image of themselves or their fellow man for another hundred years. Photography could have been invented much sooner than it was.

Though the honours for its invention are contested, and in truth claimable by no one figure, the Frenchman Joseph Nicéphore Niépce was the first to create original pictures using an optical apparatus that we would recognize as a camera and a silver chloride medium. Louis Daguerre continued his work using silvered plates sensitized with iodine vapour to produce a film of silver iodide which was then exposed to the scene to be recorded. The silver iodide was converted back to silver where the light struck it to create a negative image. Deposited directly on the silver mirror surface, however, this negative could be made to appear as a positive image simply by altering one's angle of view. Many others made important contributions, Humphry Davy, William Fox Talbot and John Herschel among them, but neither the artists, dashing between the sunlit world and the dark room, nor the chemists observing silver's abrupt transitions from white to black and black to white ever paused to consider the deeper meaning of the metal under their gaze.

## The Worldwide Web

Christopher Wren's vision for the rebuilding of London after the Great Fire of 1666 was an unashamed product of its era, a rational grand plan based on modern scientific principles that would sweep away the fetid tangle of medieval lanes that had enabled the fire to cause such devastation. But the proposed city layout was only ever realized in small part. The vistas that Wren imagined stretching from Ludgate in the west to Aldgate and the Tower in the east, and the great piazzas with their octagonally radiating streets, never materialized – such grandiose Parisian-influenced designs stank too highly of a royal absolutism unbearable so soon after the Restoration of the monarchy. At the heart of the plan, St Paul's Cathedral was rebuilt to Wren's design, and serves now as a token of the ideal city that the architect saw in his mind's eye, one that might justly have laid claim to be a modern-day Rome.

Wren studied the world's greatest domed buildings, taking inspiration from Italy as well as Byzantine and Islamic architecture, including works such as the Hagia Sophia basilica in

Constantinople, in order to devise a means of raising the largest dome that he could. Greatest of all was the concrete dome of the Pantheon in Rome, whose bronze covering was looted in 1625 by Pope Urban VIII for more pressing projects. For the weather-tight covering of London's new cathedral, Wren concluded in favour of pure copper, which could be beaten thinner than other metals to create a lightweight roof that needed fewer support-ing columns and would therefore allow a maximum quantity of light to percolate through the vast interior.

For Wren, copper had a visual and symbolic advantage as well as structural benefits. Over time, the metal would acquire a pale green patina that would make the dome the most conspicuous feature of the recreated city. Among the stone towers and spires of other churches, St Paul's would stand out as the beacon of a new age of science. However, the architect's preference for cop-per met with opposition in Parliament, as had his city plan before. Daniel Defoe, who once personally supplied Wren with building materials from his Tilbury brickworks, describes in *A Tour through England and Wales* how the discussion proceeded in true English fashion along resolutely practical lines: in response to those who thought 'the copper covering and the stone lan-thorn' too heavy for the massive columns below, Wren insisted that his structure could support not only the roof but 'seven thousand ton weight laid upon it more than was proposed'. As for himself, Defoe admired the 'unashamedly continental (and High Church) design' of Wren's dome that may well have been the real bone of contention.

Wren also wanted to see copper embellishment of the Monu-ment, the commemorative Doric pillar that he and the scientist Robert Hooke designed for the spot near St Paul's where the Great Fire had started. Apparently oblivious of the irony, the architect proposed to top the monument off with 'A Ball of Copper, 9 foot Diameter . . . by reason of the good appearance

at distance, and because one may goe up into it; & upon occasion use it for fireworkes.' But copper once again proved too revolutionary. In the end, the chosen design was based on an earlier idea favoured by the king of a 'large Ball of metall gilt'.

The dome of St Paul's was finally built with a sheath of grey lead, requiring much rethinking on Wren's part as to how the metal sheets would be attached and how their greater weight would be supported. The fact that the lead roof weighed in, by some estimates, at 600 tons heavier gives the lie to the practical argument made against Wren's preferred copper. Wren may have got his calculations right, but he seems on this occasion to have made a fatal misjudgment of English character. Three hundred years later, it is impossible to imagine this familiar landmark capped either in the metallic red of new copper or the green that would have gradually taken its place as the acid from the city's fires rained down upon it. The lead umbrella seems so right in a country characterized by grey skies that we seldom think of what might have been.

Copper did eventually find its way on to the dome of St Paul's in a small way. In 1769, Benjamin Franklin, famous for his proposal to fly a kite in a thunderstorm in order to demonstrate that lightning is electrically generated, visited Britain and personally oversaw the installation of lightning conductors on the building. These were of the kind that he advocated generally for buildings and ships, based on a long rod or bar of iron. Three years later, the cathedral was struck by lightning, and the iron was observed to glow red hot as it struggled to transmit the charge to earth, and the great cathedral was once again threatened with destruction by fire. After this, Franklin's lightning conductor was replaced with a more expensive copper one that would conduct electricity more efficiently and pose less of a fire hazard.

Copper possesses a unique portfolio of properties, identified and exploited at different periods in its long history. Between them, they have ensured that the element has never slipped from its prime position since it began to be worked by man more than 6,000 years ago. The most immediately striking of these is of course its colour. It is the only red metal. This gave copper a special status in relation to gold, the only other coloured metal. In the New World, European explorers such as Cabot in the north and Cortés in the south found the metal used for jewellery and devotional purposes. The Florentine navigator Giovanni da Verrazzano believed copper was 'esteemed more than gold' by the natives. The contrast of colour between the pure red metal and its watery blue and green salts was widely felt to be significant too. This embodiment of opposites was regarded as symbolic in cultures as diverse as the Aztecs and the Dogon of Mali, for whom the accretion of green corrosion on the brown metal symbolized the return of vegetation after rain.

The first of copper's useful properties to be exploited was its malleability. It was soft enough to be hammered and beaten into useful artefacts yet hard enough that these items would be serviceable. The ancient Egyptians used copper to make swords and helmets and even drain pipes. Being abundant as well as malleable, copper was more practical for coins than gold and silver, but sometimes raised objections from the people among whom it was circulated because of the obvious disparity between its face value and its actual worth. Henry VIII came to be known as Old Coppernose because he introduced so much copper into the silver coin of the realm that raised parts, such as the king's nose, would turn red as they wore down. Later innovations meant that copper could be machine-rolled into thin sheets, producing the now familiar roofing material used for the domes of European cathedrals and, in due course, the new capitol buildings of North America.

The metal's ready conduction of both heat and electricity were the next properties to be recognized. The American patriot Paul Revere achieved fame with his copper-bottomed cooking pots and pans at the beginning of the nineteenth century. At the same time, scientists investigating electricity found that copper would carry an electric current better than any other material apart from silver. Alessandro Volta made his first electric pile from layers of zinc and silver, but thereafter most batteries used copper.

But it is one final property, its ductility, that has given copper its greatest role in transforming our world. It is the fact that copper can not only be beaten into a sheet but also drawn into a wire, a wire furthermore that conducts electricity, that led to the creation of what can fairly be described as the first worldwide web.

The cabling of the world relied on a number of key breakthroughs made in a relatively short space of time – batteries that could deliver a steady current; galvanometers that could detect an electrical signal and show it by the deflection of a needle; copper refined to sufficiently high purity to conduct electricity efficiently; and the discovery of the insulating properties of gutta-percha, a resinous rubber-like substance obtained from Malayan sapodilla trees.

The first primitive electric telegraph line was built in the 1790s by Francisco Salva and was capable of transmitting sparks from Madrid to Aranjuez fifty kilometres away. Salva proposed a separate wire for each letter of the alphabet with the arriving spark briefly illuminating letters in turn in order to spell out messages. (He apparently also considered connecting a person to each wire and having them shout out the letter when they received an electric shock.) Many equally eccentric schemes were essayed in the years that followed, spurred on by the obvious need for a more effective means of communication than the visual methods of flags and lights used by relays of signalmen during the Napoleonic Wars. But efforts were held back by a poor basic understanding of electrical phenomena. It was not until 1831, when Michael Faraday first wound copper wire round an iron ring to demonstrate electromagnetic induction, that the relationship between various kinds of electricity and conducting matter was better understood.

Charles Wheatstone and William Fothergill Cooke demonstrated a more practical telegraph in 1837 when they built a two-kilometre connection along the railway between Euston and Chalk Farm in London, itself then newly laid. A similar trial connection established on the Great Western Railway between Paddington and West Drayton two years later was extended to Slough in 1843. This telegraph caught the public imagination soon after its installation when John Tawell, having murdered a woman in the town, boarded a London-bound train thinking to make his escape. He failed to allow for quick-witted station staff who telegraphed ahead. The police arrested him when he duly disembarked at Paddington.

In 1838, meanwhile, the American inventor Samuel Morse was in England seeking a patent for his own telegraph system. Wheatstone used his connections to ensure that his competitor's application was rejected, and Morse had to content himself with a seat in Westminster Abbey, where he witnessed the coronation of Queen Victoria before returning to the United States to obtain a patent there for the coded telegraphic method that still bears his name.

Progress was rapid from these modest beginnings as the inventors set themselves the target of bridging successively wider chasms. The greatest challenges were the same as they would be for powered flight fifty years later: first the English Channel, and then the Atlantic Ocean. Undersea cables posed far greater challenges than land lines, which could simply be buried or carried overhead on poles. The wire had to be prefabricated in long lengths and wound on to rolls so that it could be spooled out at sea from specially adapted ships. In 1850, Jacob and John Watkins Brett successfully laid a gutta-percha-insulated copper cable between Dover and Calais, but the connection broke after a day. According to one story, the fisherman who dredged up the severed cable saw the gleaming metal at its core and thought he had struck gold. A cable of four independently insulated wires, protected by layers of hemp and tar and strengthened with iron wire, that was laid the following year proved more durable. In the following decade, England was joined to Ireland, Denmark to Sweden, and Italy to Africa via Corsica. Newfoundland was linked to Nova Scotia across the Cabot Strait and thence overland to New Brunswick, Maine and the rest of North America. All that remained now in order to complete the cable connection of Europe and America was to make the link from Ireland to Newfoundland across nearly 2,000 miles of the Atlantic Ocean.

The technical requirements for this far longer and deeper undersea connection were colossal. There was no possibility of boosting the signal at intermediate points along the cable, as there was with cables on land, so the copper conductor had to function as a single length. This made it critical that the engineers minimized signal losses due to resistance in the wire and the effects of immersion in seawater, itself a highly conductive medium. The Scottish physicist William Thomson, later Lord Kelvin, who was appointed as a scientific consultant to the Atlantic Telegraph Company, delighted in a problem that allowed him to deploy his mastery of the new theories of electromagnetism for a practical end. He wrote to his friend Hermann von Helmholtz:

It is the most beautiful subject possible for mathematical analysis. No unsatisfactory approximations are required; and every practical detail, such as imperfect insulation, resistance in the exciting and receiving instruments, differences between the insulating power of gutta-percha and the coating of tow and pitch round it . . . gives a new problem with some interesting mathematical peculiarity.

Thomson advocated the use of a thick copper wire, running small currents through it which could be picked up with sensitive detectors, but he was outmanoeuvred by company men who favoured the cheaper option of pushing stronger signals through narrower-gauge wire.

The first attempt to make the crossing was set for the summer of 1857, the same year that the world's largest copper-domed roof was completed over the Reading Room at the British Museum. In August, the massive HMS *Agamemnon* and the United States frigate *Niagara* accompanied by a flotilla of support vessels set sail from Valentia on the west coast of Ireland carrying 1,200 two-mile lengths of copper wire, pre-joined into eight 300-mile lengths. The cable weighed about a ton per nautical mile, most of which was accounted for by the outer strengthening of steel wires and insulation; the copper made up only 107 pounds per mile of the overall weight in wire no thicker than a pencil lead.

As the final plans were laid for the voyage, Thomson made a crucial further discovery that the purity of the copper greatly affected its conductivity. Practically his last action before boarding ship was to read a paper to the Royal Society, 'On the Electrical Conductivity of Commercial Copper of Various Kinds', in which he revealed his important new findings. Nobody had given this matter any consideration. Despite his scientific misgivings, Thomson dutifully sailed aboard the *Agamemnon* in his capacity as a director of the Atlantic Telegraph Company, while Samuel Morse battled with seasickness and a leg injury aboard the *Niagara*.

It might never have worked properly in any case, but only 400 miles out from Valentia, the cable broke, and the task was given up for the winter. The following summer, two more attempts were made to complete the job using the same ships and picking up the same cable. The first was defeated by unseasonal gales. The second attempt seemed successful, but celebrations proved premature when the connection was lost after less than a month. Recriminations were followed by an inquest, which showed that the cable had been fatally damaged by attempts to boost the signal strength through the application of higher voltages than those for which it had been designed – exactly the mishap that William Thomson had feared.

Anglo-American relations deteriorated during the American Civil War to the extent that President Lincoln preferred to make overtures to Tsar Alexander II, suggesting a cable from Alaska to Siberia and through Russia to the cities of Europe, rather than persist with the Atlantic project. However, a permanent transatlantic cable was finally laid in 1866 by Brunel's steamship *Great Eastern*. A correspondent from *The Times* likened her to 'an elephant stretching a cobweb'. In addition, a cable abandoned the previous year was also finished, providing a spare line and reassurance to sorely tried shareholders in the telegraphic enterprise that this time the connection really would last. The design of these cables had been modified along the lines that Thomson had previously proposed, using three times the amount of copper in seven wires – 365 tons in all – with every length tested in advance for purity and conduction.

After the cable went operational, one of the engineers performed a simple test on the line at Valentia. He cabled to ask that the two lines be electrically connected at the Newfoundland end, and proceeded to make a little electrolytic cell using a piece of zinc and a splash of acid in a thimble. The zinc was then connected to one copper end of the cable while the other copper end was dipped in the acid. The single volt generated by this makeshift battery was enough to drive a current 3,700 miles across the ocean and back.

Further cables across the Atlantic and elsewhere were swift to follow, backed by many national governments, while Britain sought to link all of its dominions. In 1901, at the end of Queen Victoria's imperial reign, the Cable Steamship *Britannia* laid parts of a cable across the Pacific Ocean from Australia and New Zealand via Norfolk Island, Fiji and the remote Fanning Island to Vancouver – completing the union of the pink nations on the world map by pink copper.

The world today is cocooned in copper wire and, notwithstanding the advent of optical fibres and satellites and wi-fi, more than half the copper mined is still drawn into wire or otherwise employed in communications and electrical applications.

Though largely hidden from view, copper has become the symbol of civilization that Wren believed it to be when he thought to use it to cover the dome of St Paul's.

## Au Zinc

Nobody has put their stamp on the city of Berlin more thoroughly than the Prussian architect Karl Friedrich Schinkel. Though he could do Gothic if asked, he is celebrated for his development of a Greek-influenced neoclassical style that tempers its monumentality with superb detailing. It was in this idiom that he designed many of the cultural edifices that give Berlin its austere grandeur today – the Schauspielhaus, the Altes Museum, the Singakademie – as well as churches and villas, and buildings for his patrons, King Frederick William III and his heir, at nearby Potsdam.

These buildings are forthright and impressive, as they were required to be in order to express Prussia's recently regained independence from Napoleon's armies and the accompanying influence of the fussy French Beaux Arts style. But appearances are sometimes deceiving. Schinkel began his career designing theatre sets – he devised the still famous hemispherical backdrop of stars for a production of *The Magic Flute* – and was sometimes more concerned to achieve the right effect than with authenticity. Thus, the statues that punctuate the cornices and pediments of his buildings are not always the stone or bronze they appear to be but are in fact sometimes hollow zinc. Schinkel also designed the Iron Cross, Germany's highest military decoration, but, belying its name, even this medal was sometimes partly made of zinc.

Zinc was the first useful metallic element to come to light since iron, lead and tin, discovered thousands of years before. A thirteenth-century Indian text describes how the metal was made by heating calamine, a traditional medicament which is principally zinc oxide, with organic matter. This makes zinc the only element with an attributable date of discovery where Western science cannot claim to have made the running. News of the metal came to Europe via China, which was first to exploit zinc on a large scale. The renowned alchemist Paracelsus reported rumours of the new metal in the sixteenth century, and, not long after, samples of zinc wares were brought to the West on trading ships. It was not until the eighteenth century that deposits of ore were located that enabled the metal to be smelted in Europe.

Zinc inhabits a no-man's-land between the metals of antiquity and the modern metals teased from possessive ores by the ingenuity of science and the might of the Industrial Revolution. Its ambiguous position is underlined by the fact that it has been used unwittingly for thousands of years in the form of brass (an alloy of copper and zinc, known long before zinc itself because their ores often occur together). Zinc would be quick to find application in its own right, but emerging as it did in this roundabout way, it had none of the cultural baggage that copper clearly had for Christopher Wren.

To Schinkel, this blank history spoke of possibility. The architect became the champion of the zinc foundries that sprang up in the 1830s, using the metal for the statues and decorations on some of his later buildings, and exhorting other architects to do the same. Often stamped from sheets of the metal rather than cast, the 'white bronze' quickly became popular for statuary of all sorts, especially where weight or cost ruled out real bronze. It was not long before popular zinc figures of cemetery angels and garden deities were being pressed out daily. The trend spread to the United States when one Moritz Seelig fled the 1848 revolution in Germany to set up a zinc foundry in Brooklyn. He

prospered as, like Schinkel in Berlin, mayors across America sought to embellish their towns with the grandest possible sculptural figures purchased on the smallest possible budgets. The statues of Justice and the Civil War memorials that now crumble gently in the parks and plazas of provincial American towns were largely selected from Seelig's trade catalogues.

Zinc has found a market in architecture but perhaps not yet a role. However, one remarkable Berlin building may change this.

The commission for a new Jewish Museum in the German capital (a previous one had opened in 1933, three months before Hitler came to power) was won by Daniel Libeskind in 1989. Of the 165 competition entries, the young American's design, based on the fragmented music of Schönberg, the writings of Walter Benjamin and motifs inspired by other Jewish intellectuals responsible for enriching German cultural life, struck the jury as the most brilliant and most complex – if possibly unbuildable. It proved quite buildable, though, and upon completion in 1999 was judged so remarkable a structure that it was opened to the public even before any exhibits were installed. Visitors paid to experience tunnel-like voids and twisted, ever-receding spaces that seemed to manipulate perspective and even gravity itself, producing effects as unsettling as any material environment can offer.

The exterior is hardly less disconcerting. The building describes a jagged squiggle on the ground with sheer walls rising on all sides entirely sheathed in parallelograms of zinc. Strip

windows slice diagonally across this façade and each other at apparently random angles tracing out what may be a deconstructed star of David or a zigzag path of wandering and loss.

Libeskind has explained that he chose zinc in response to Schinkel's call, and as an obvious gesture of harmonization with the adjoining Berlin Museum, whose windows are framed in zinc. But there is deeper symbolism that makes the material specially appropriate here. In dream interpretations, I discover, zinc is associated with migration. It is a natural choice for a building celebrating a citizenry of emigrants who have emigrated once more. This symbolism may perhaps be explained by zinc's poor historical timing, arriving too late to find a partner in the alchemical dance that paired the metals with the bodies of the solar system. Copper, iron, tin and lead are each associated (somewhat differently, according to whose tradition you're following) with a planet. But zinc dances alone. Zinc is also said to symbolize progress towards a goal, which seems apt for a building that is, in Libeskind's words, 'always on the verge of becoming'.

More obvious is zinc's connection with ceremonies of preservation and burial. When William Deedes, the journalist who provides the model for the central character in Evelyn Waugh's *Scoop*, was sent to cover the Abyssinian War, he travelled with his possessions in a cedarwood trunk that was lined with zinc in order to keep out the ants. The metal is also often used to line coffins as a relatively cheap, safe alternative to lead: my chemical informant Andrea Sella has a powerful sensory memory from his Italian boyhood of burial preparations being accompanied by the sound of a blowtorch as it's used to seal the zinc of the coffin before the lid is screwed shut. The German artist Joseph Beuys has employed zinc chests in some of his works as containers for fat. Although it is the fat that has received most of the critical attention, recognized along with felt as one of Beuys's trademark materials, the zinc is significant too, chosen not least for its representation of opposites: as poison and salve, as a seal that ultimately crumbles. In this context, Libeskind's building becomes a vast sarcophagus, a metaphorical container of the bodies of the six million Jews murdered in the Holocaust, as well as a means of preserving their memory.

Zinc is also used for the hygienic transportation of bodies across national borders. The metal provides a two-way barrier. It is there to prevent the ingress of contamination that would hasten the body's decay, but it also serves to seal in potentially infective matter. In a poem by Bertolt Brecht, 'Burial of the Agitator in a Zinc Coffin', it is also an impermeable layer that preserves a sinister mystery. The poem was set to music, along with another, 'To the Fighters in the Concentration Camps', by Hanns Eisler, a pupil of Schönberg, in his colossal *German Symphony*. The piece was intended to be performed at a music festival connected with the 1937 Paris Universal Exposition, but Nazi pressure forced the organizers to propose that the vocal parts be replaced by saxophones so that Brecht's words would not be broadcast. Eisler naturally refused this suggestion and substituted an earlier composition in the festival programme. The *German Symphony* was heard for the first time only in 1959. Brecht's poem begins: 'Hier, in diesem Zink / liegt ein toter Mensch . . .':

> Here in this zinc
> lies a dead person,
> or else his leg and his head,
> or still less of him,
> or nothing at all since he was
> an agitator.

Paris has happier associations with zinc. Everywhere I turn I see rooftops made from pale sheets bent into curved mansards. At some point the material must have taken over from lead and slate, with the pleasing consequence that the roofs are no longer dark lids on top of the buildings, but dissolve effortlessly into the milky blue sky.

At night, however, it is in the bars that the metal is supposed to be found. The English language has its share of elemental synecdoche – we use irons, we spend nickels and coppers, we once made carbons of important documents. But in Paris in its early twentieth-century heyday the bars became *zincs*. Jacques Prévert put the drunken ramblings of a *zingueur*, as the city's zinc roofers were known, at a zinc bar into a poem, and Yves Montand turned it into a famous chanson, 'Et la fête continue'. I find one of the few remaining zincs on the Left Bank just round the corner from the celebrated Deux Magots and the Café de Flore. Perhaps Ernest Hemingway and Gertrude Stein once propped up this bar too. Now operated by a restaurant chain, the place knows its zinciness is its pedigree and flaunts it accordingly. The chairs are coated with metallic paint, the restaurant name cut out of sheet metal, the menus dressed in grey. The bones of some extravagant Art Nouveau ironwork still hold up the building. But of the authentic *zinc* there remains less than a barman's arm's length, now taken over for the maître d's lectern, with its intricate bas relief of grapes and vine-leaves wrought in the dull grey metal. Across the room is a shiny new bar, but it gleams suspiciously brightly in the tones of another metal.

Puzzled by this, I tracked down the only artisan still supplying and restoring these bars. At the Ateliers Nectoux round the back of La Défense, the suburban business district, Thierry Nectoux reveals that all his work is in fact done in tin, as it has been for three generations. 'There's never been zinc in the studio,' he tells me. 'Zinc can't be put on countertops because it is not *alimentaire* [suitable for use with food], and it oxidizes. Also, it's not easy to cut cold or to work or clean. Tin is completely the opposite.' I could see some sense in this. Everybody remembers from school chemistry that zinc dissolves in acid – it would not get on well with spilt lemon juice or Coca-Cola.

But if they are made of tin how did the bars come to be called zincs? Nectoux's thoughts on the matter seem fanciful. One of his suggestions is that they acquired their name from the *zingueurs* who would drop in to these bars for a dose of vertigo-defying Dutch courage before work. This doesn't sound right. Surely, zinc bars were called zincs because they were once truly made of zinc, and tin was an adulteration of this tradition. My French-speaking grandfather's *Larousse de Poche* seems to confirm this hunch. Published in 1922 at the height of the zinc era, the dictionary acknowledges the colloquial meaning of the word as a counter over which wines are sold. It does not elaborate as to its origin, but nor does it say anything to suggest the bars weren't genuinely made of zinc.

## Banalization

The waters had begun to rise decades before, but the surge tide of literary modernism really broke in 1922 with the publication of *Ulysses* and *The Waste Land*. That year also saw the first performance in a Bloomsbury drawing room of a musical entertainment called *Façade*: music by the twenty-year-old composer William Walton lolloped along to Dadaesque words by Edith Sitwell, the poet and doyenne of English eccentrics, who enunciated her part through a megaphone from behind a curtain. The twenty-odd auditors of the private recital were variously baffled and exhilarated. The public premiere the following year was met, predictably enough, with general ridicule. It was during this period of wild experiment that Edith's younger brother Osbert commissioned a sculpture of his sister from another member of their set, Maurice Lambert. Castings of the head, somewhat less than life size, now reside at Renishaw Hall, the Sitwell ancestral home in Derbyshire, and in the National Portrait Gallery in London. The head itself is small and oval, supported by an elongated, gently curving neck. The modish angular crop of the hair and sharp nose give the work a perhaps not unintended resemblance to a Saxon helmet. But any primitivism is offset by the material: the heads are cast in aluminium.

Neither the Sitwells of the present generation nor Edith's biographer knows who chose aluminium. Nor does the biographer of Maurice and his composer brother Constant Lambert. When Maurice Lambert sculpted the head of Walton a couple of years later, it was conventionally made in bronze, from which we can infer that aluminium is more likely to have been Edith's idea than Lambert's. Suffice it to say that the choice of material inadvertently reflected the majority critical opinion that Edith Sitwell's artistic project was both lightweight and quite unnecessarily modern.

In Britain, you almost had to be an eccentric to see merit in aluminium. It was left to nations less ambivalent in their attitude to technological novelty to find ways of putting the metal to more functional use. While the British fought the class war with their silver and their pewter, the French and the Americans turned aluminium into objects that were swiftly recognized as icons of progress and modernity – things on legs, such as the furniture of Charlotte Perriand and Charles Eames; things on wheels, such as the Airstream trailer and the first Citroëns 2CV. Aluminium cut the ties to the past and brought new hope of mobility and liberation. The Greyhound bus, with its signature ribbed aluminium trim and its blatant promise of freedom, was the creation of a French émigré to New York, the flamboyant industrial designer Raymond Loewy.

Long before it could attain this popular appeal, aluminium enjoyed a brief spell of imperial patronage. This now ubiquitous material – as vital to us as steel and more visible than any of the metals known in antiquity – was only isolated as recently as the 1820s, and it was not until the 1850s that an even remotely commercial way was found to separate it from its ore, bauxite, named after Les Baux in Provence, where it is still possible to see the bleached quarry works on the hill above the town. The process developed by Henri Sainte-Claire Deville in Paris involved heating compounds of aluminium with sodium metal, which was itself exceptionally hard to obtain, and this made his aluminium hugely expensive. Though it scarcely seems credible now, aluminium was hailed as a new precious metal to be placed along with gold and silver – its sheer cost and exoticism compensating for its low density and diffuse shine – and it was worked and flaunted in ways that reflected this status.

Deville achieved his breakthrough at an opportune moment. Paris was excited by rumours of the new 'silver from clay'. Deville presented a clutch of little ingots of aluminium for the first time at the Paris Universal Exposition of 1855, where they were admired by the Emperor Napoleon III, who promptly gave his financial encouragement to the chemist. The metal was then priced at 3,000 francs a kilogramme and worth a dozen times its weight in silver. But this was if anything an incentive rather than a deterrent to the greatest craftsmen of the day. The renowned goldsmith Christofle became interested in the new material and made some of the first handcrafted items of tableware and jewellery. The emperor is said to have given banquets at which the most honoured guests were given aluminium cutlery while the hoi polloi had to make do with silver and gold. Napoleon's son and heir, the Prince Imperial Eugène, born in 1856, was given an aluminium rattle, a clear signal that the country should embrace the new. The brass eagles adorning the flagstaffs of the imperial guard were recast in aluminium, a gesture presumably appreciated by their bearers. Although craftsmen like Christofle exploited the metal for ornamental purposes mainly because it was regarded as precious, Napoleon saw that it was aluminium's lightness that could be its most valuable property. We may see hints of this promising future in a few objects made at this time that straddle the worlds of function and ornament, such as medals and opera glasses. But at the height of the Industrial Revolution, with iron the engineering wonder of the day, it is no surprise that the greater potential of aluminium was not more widely recognized.

In *The Theory of the Leisure Class*, Thorstein Veblen chooses an aluminium and a silver spoon to illustrate his dictum that the utility of objects valued for their beauty 'depends closely upon the expensiveness of the articles'. The utility Veblen refers to is social rather than functional; he's saying we tend to value things more highly when we know they are expensive. By the 1890s, when Veblen was writing the work that would give us the expression 'conspicuous consumption', aluminium was cheap; the aluminium spoon might cost ten or twenty cents, the silver spoon as many dollars. We know the lighter aluminium spoon is easier to use, yet we prefer the silver because it 'gratifies our taste'. The insubstantial weight, machine manufacture and general plainness all betray the aluminium spoon as the one we should spurn.

In 1855, though, Napoleon III's patronage entirely reversed the situation. For a brief moment in the unlikely halls of the Louvre Palace, it was aluminium that was skilfully wrought, caressed for its lightness and admired for its mysterious pallor. The emperor did not want this state of affairs to persist any longer than it had to, however. He was fired by the idea that the new metal could be used to make armour and weapons, and in 1856 a prototype aluminium helmet was set before the French Academy of Sciences. The assembled sages judged it robust and serviceable – and, a little beside the point, beautiful too. But reluctantly they had to report that it was also far too expensive. It would be nearly a century before Napoleon's hopes for aluminium as a utilitarian metal would be fulfilled.

The United States Congress 'nearly put a shiny lining of aluminium foil atop the Washington Monument', according to Bill

Bryson's *Short History of Nearly Everything*, 'to show what a classy and prosperous nation we had become'. In fact, the monument *is* capped with aluminium, although the job was done with none of the symbolic intent that Bryson implies. It took some doing. Congress set the ball rolling in 1783, at first merely giving its approval for an equestrian statue of the general who had led them to independence. Six years later, George Washington became the nation's first president, serving in that office for eight years. By the time Washington died in 1799, the city named after him was growing into its pomp. The Capitol building was rising, the first pearl in a necklace of neoclassical temples to democracy, and the thought crept in that something more majestic was required to honour the father of the country. The cornerstone for the colossal marble obelisk we see today at the axial intersection of the Mall was eventually laid in 1848, and the monument finally dedicated in 1885.

The topmost twenty-two centimetres of what was then the tallest manmade structure in the world were made up by a lightning-conducting pyramid of cast aluminium, its point as sharp as a pencil. Various metals had been considered, including copper, bronze and brass, which would then be plated with platinum. Colonel Thomas Casey of the US Army Corps of Engineers chose aluminium 'because of its whiteness and the probability that its polished surfaces would not tarnish upon exposure to air'. A spider's web of copper lightning rods runs hidden from view from the aluminium point down to the ground. Although there was no declared attempt to send any form of cultural signal by setting aluminium in this beacon position, the moment lives in memory, and especially in the corporate memory of the American aluminium industry, which still trades on its Washington connection. At a dollar an ounce, aluminium by now cost about the same as silver – as luck would have it, the price began to tumble as soon as the monument was completed. In December 1884, though, when the little aluminium pyramid was briefly displayed to the public in New York before its final installation, it was still definitely regarded as a precious metal: the exhibition was at Tiffany's, the famous Fifth Avenue jeweller. Christmas shoppers took it in turns to leapfrog the futuristic menhir that would soon soar higher in the sky than any other artefact of man.

These shiny embellishments may be functional but, purposely so or not, they are also rhetorical pronouncements from the centre of government. Both Napoleon III's cutlery and Washington's monument are explicit tokens of commitment to modernity by the state. Other elements, such as neon and chromium, as we shall see later, became signs of aspiration and hope for the future, but these were to be popular enthusiasms, cheap, cheerful and democratically spread through the land. Aluminium was a plaything and a project of leaders. But it would not remain so for long.

The history of aluminium is 'a process of banalization', according to the company history of the state producer L'Aluminium Français. During a single century, aluminium usage has travelled from the singular to the general and from the general to the banal – a progress that iron and copper completed over millennia. The greatest step in this journey was the one necessary for it to begin at all, the one that was to knock aluminium off its pedestal as a precious metal. Fittingly, the breakthrough was achieved simultaneously by a Frenchman and an American. Paul Héroult and Charles Martin Hall were both in their early twenties in 1886 when they separately perfected a process that used an electric current instead of the chemical power of sodium to release aluminium from its ore. The metal is still made electrolytically today. As the price of aluminium fell far below that of silver, and eventually below even copper, makers like Christofle lost interest in it, and it could begin to fulfil its true destiny as the new industrial wonder metal. The sophisticated method of making it underlined its credentials as something utterly modern: linked umbilically to the 'second industrial revolution' triggered by the wide availability of electricity, aluminium was set to be the very embodiment of the technological twentieth century.

America and France may have pioneered the development of aluminium, but they disagreed over its spelling. Even the great editor H. L. Mencken is at a loss to explain this. In *The American Language* he is forced to confess: 'How *aluminium*, in America, lost its fourth syllable I have been unable to determine, but all American authorities now make it *aluminum* and all English authorities stick to *aluminium*.' Other sources suggest it may have been the doing of Charles Hall. The patents he took out for his electrolytic refining process referred to 'aluminium' while his commercial publicity material touted the merits of 'aluminum', whether by intent or typographical error is not known. The shorter word spread and stuck in the United States; in France, Britain and the rest of Europe, the extra syllable remained.

But perhaps the boot should be on the other foot. Rather than asking how the name was shortened, we should go in search of the British fusspot who insisted on the extra syllable in the first place. Humphry Davy, who repeatedly tried to isolate the metal, himself christened it aluminum straightforwardly enough after its ore alumina (an improvement on his first thought, alumium). But then in 1812, an anonymous reviewer of Davy's *Elements of Chemical Philosophy* writing in the *Quarterly Review* objected to the 'less classical sound' of this word and, conveniently forgetting the precedents of platinum, molybdenum and the recently named tantalum, cast the die for aluminium as being in tune with the many other elements terminating in -ium.

★

The Hall–Héroult process provided the spark. Aluminium, the most abundant metallic element in the earth's crust, could now be put into the service of man, thanks to the rapidly expanding power of electricity. The most visible early applications were in transport, where the metal's low weight was a great advantage. The French car manufacturers Renault and Citroën, always famous for their innovative designs, investigated aluminium thoroughly in the 1920s. They used it at first not to replace heavy steel panels but for wheels and decorative features such as hubcaps (which the French charmingly call *enjoliveurs*, meaning prettifiers). The metal was employed on a larger scale to sheath industrial machinery and on custom-built transport such as railway carriages and delivery vans. A Pullman railway coach displayed at the Chicago Century of Progress Exhibition in 1933 weighed just half what a standard steel coach weighed. The Paris Universal Exposition in 1937 featured an Aluminium Pavilion, and the metal was copiously incorporated into the Alexander III and Alma bridges and elsewhere around the City of Light.

But it was only when sheets of the stuff were pressed and bent into seductive curves for aerodynamic performance that the romance of it was truly unleashed. Aluminium was taken up more widely for the skin and skeletal structure of passenger aircraft quite suddenly in 1931 after the fatal crash of a wooden-framed airliner carrying a famous football coach to Los Angeles. Aircraft such as the Douglas DC-3, the glamorous transport of Hollywood stars, inspired terrestrial imitations in the form of cars and buses and mobile homes whose lustrous, bulging forms provided the glimmer of a better life after the Depression. The decor at the Radio City Music Hall in New York is dominated by curving horizontal bands of aluminium. But the Airstream trailer went farthest in reaching for the skies, aping even the rivet lines of aircraft panels along its contoured skin. At a time when trailers in Europe looked, in the view of one design critic, more like gypsy caravans 'lacking only a thatched roof', this American design, developed with help from one of the creators of the *Spirit of St Louis*, the aeroplane in which Charles Lindbergh had flown from New York to Paris in 1927, revelled in its buxom aluminium nakedness.

Aluminium moved swiftly into the home, taken up with enthusiasm by industrial designers and by housewives, who appreciated its low weight and the fact that it didn't need polishing. The metal could be treated in new ways which added to its modernistic appeal. The most representative treatment, made famous by the designs of Russel Wright, was spin-casting, in which the molten metal was poured into a rotating mould. Mary McCarthy went so far as to praise Wright by name and 'the wonderful new spun aluminium' in her novel *The Group*. Aluminium took over in kitchen utensils that would formerly have been made of pewter. Because it retained heat better than copper or cast iron, it was also perfect for 'stove-to-table' pots and pans, a boon in servantless households. Everywhere, the smooth round forms, frequently accentuated by the horizontal lines of the spinning process and a final brushing, spoke of the streamlined new order of things.

After the Second World War, new demand was matched by new capacity, and aluminium began to be considered for the construction of entire homes. At Wichita in Kansas, the visionary designer-poet Richard Buckminster Fuller turned over an entire aircraft factory to producing his domed aluminium houses, an achievement he celebrated by penning this ditty:

> Roam home to a dome
> Where Georgian and Gothic once stood
> Now chemical bonds alone guard our blondes
> And even the plumbing looks good

Based on a circular plan, Fuller's houses looked like habitable versions of the vessels designed by Russel Wright a decade before. In France, Jean Prouvé, a pioneer architect in metal prefabrication, used aluminium panels in emergency housing for people made homeless by the war, and went on to devise flat-pack metal homes for the last generation of colonial over-seers in French West Africa. Even the British built thousands of aluminium-panelled houses in the 1940s, though they were joyless huts in comparison with the stylish prototypes of the French and Americans.

Fuller's round houses never caught on, but the aluminium of which they were made was too cheap and practical to ignore. The unstylish legacy of this bold post-war experiment rests in thousands of acres of corrugated aluminum siding that was sold door-to-door during the 1950s and 1960s, and clamped to Ameri-can homes as the latest thing in weather protection, at least until it was superseded by the next such fad, vinyl. The fictional esca-pades of two such salesmen are the subject of the 1987 film *Tin Men*. That the metal they were selling, so recently the prize of emperors, was now disparaged as mere tin is a sure sign that the process of 'banalization' was complete.

The journey from ploughshares to swords and back to plough-shares is unique to aluminium, which has a high scrap value compared with other common metals because its electrolytic extraction from bauxite is so energy-intensive. Just as Napoleon III dreamt that his aluminium cutlery might be converted into battle gear, so Lord Beaverbrook appealed through his news-paper empire for the British people to hand in their aluminium utensils to be 'turned into Spitfires and Hurricanes'. After the war, the priorities were abruptly reversed, and the catalogue of the 1946 exhibition, *Britain Can Make It*, explained how war-time production methods would lead the country back from 'Spitfires to saucepans'.

Perhaps this did happen, although for the most part people

were unaware of it. At an antiques fair in Dorset, I spotted a Picquot Ware tea-set made in the 1950s from 'Magnailium lustre', and bought it. It was unused, and the metal had an unusual lilac sheen. But what was 'magnailium' (apart from another metal with an apparently superfluous *i*)? The vendor had suggested to me that the set was made from melted-down wartime aircraft parts. I liked the way the design seemed to capture aluminium's downward spiral into domesticity from higher callings. The word 'magnailium' was presumably a composite of aluminium and magnesium. The latter metal being two-thirds as dense, the two were combined during the war to make an alloy that was lighter and stronger, if considerably more expensive, than pure aluminium.

But I had my doubts. For a start, the pieces seemed rather heavy, even allowing for the thick-walled castings. And then there was the label: 'Designed by Jean Picquot. Fashioned by craftsmen'. Who was this designer I had never heard of, and who

was not listed in the usual design references? He or she soon turned out to be the imaginary friend of the stolidly English-sounding manufacturer of the wares, Burrage & Boyde, a figment of suavity presumably conjured up in order to capitalize on the reputation aluminium had acquired in the hands of the innovative French.

By now I was thoroughly sceptical. I decided a simple test was needed to resolve the mystery of magnailium. Choosing the milk jug, the only piece of the set without a wooden handle, I first weighed it and then immersed it in water, using the displacement to estimate the volume of metal. Dividing one by the other would give me the material's density, which would be an important clue as to the metals it was made from. The density came out at around 3.9, more than double magnesium's 1.7 and greater even than aluminium alone at 2.7. My magnailium was clearly no fancy aerospace alloy. It had to be aluminium combined with a *heavier* metal such as a standard alloy with copper. I preferred the myth, however, and comforted myself with the thought that at least a few of the metal atoms in my tea-set may have flown in the Battle of Britain.

## 'Turn'd to barnacles'

When it was built, the American presidential residence in Washington, DC, was coated in a damp-repellent mixture of slaked lime and glue, and people started to call it the White House. Tombs were likewise brushed with lime to protect them from the ravages of the weather. Whited sepulchres – were ever two words as lost without one another as 'whited' and 'sepulchre'? – occur in St Matthew's gospel as an image of hypocrisy, and refer to those tombs 'which indeed appear beautiful outwardly, but inside are full of dead men's bones and all uncleanness'.

Whiteness is freedom from colour and an escape from the rainbow chaos of life. Lime's whiteness is a scourging simplicity, the purity of an ideal, the finality of a death. Whiting is the action of adding a layer of lime-wash, yet it is also a subtraction, a gesture towards liberation, a brushing away of earth and the earthly, a disencumbrance, a literal lightening and also the lightening of a load. The cleansing and preserving action of whitewashing ritually repeats the throwing of lime into the grave with the corpse. Our bodies decay, our bones are left, picked clean and bleached of all colour. We fade to white.

Lime is oxide of calcium. It is made simply by heating chalk, limestone or sea shells to drive off carbon dioxide. The strongly alkaline white powder that results then slowly absorbs water and carbon dioxide from the air, these irresistible actions being the key to its many long-established applications. Lime is used in burials because of this hygroscopic property: it draws out moisture from the body and reduces the risk of disease from putrefaction. Saturated with water, or slaked, it becomes whitewash. Lime in mortar quickly dries, replacing the water it loses with carbon dioxide, causing the soft white powder to turn to durable stone. So central was this action to the routines of life and death that lime, the Romans' calx, gave its name to the alchemists' and early chemists' generic term for burning in air or roasting, calcination. Lavoisier gave lime a place on his list of the elements, ranking it as one of the 'salifiable simple earthy substances', even as he guarded his hunch that the white substance was not itself a pure element but was hiding within it a new metal that science was as yet unable to extract. Calcium was only prised free from its indispensable oxide in 1808 when Humphry Davy subjected it to the electrolysis he had already employed in the discovery of potassium and sodium. The metal was not made on a large scale for another 100 years.

Calcium, then, is the element at the chemical heart of lime, limestone, chalk and many minerals besides, such as calcite and gypsum. Calcium may not be the only element to form predominantly or entirely white compounds, but through these important and abundant natural materials it is the element we most associate with the absence of colour. Apart from snow, our similes for whiteness are calcareous – white as marble, alabaster, chalk; white as ivory, bone or teeth; white as pearl. Calcium's whiteness is iconic: I hesitate to use such an overworked adjective but the instance of the White House alone seems to sanction it. The White Cliffs of Dover, too, were a potent enough image that the American lyricist of Vera Lynn's wartime song felt able to complete his work having no more seen them with his own eyes than he'd seen the 'bluebirds over'. The white horses and other figures created by carving out turf from England's chalk hillsides in Neolithic times – with occasional additions still being made today – also retain a timeless graphic power.

Walking the hills and downs of southern England today one can still feel how a collective identity can spring from autochthonous rock. Simply drawn on paper, figures such as the Cerne Abbas Giant or the White Horse at Uffington would appear as nothing more than graffiti, vulgarly priapic and Picasso-ish respectively. But inscribed in chalk, they become English despite themselves. On the Isle of Wight where horizontal geology is tipped up on edge like a slice of layer cake, I head to the western tip and the standing chalk rocks known as the Needles – once four, now three plus their vital lighthouse, and never quite as craggy in fact as in the old prints etched by artists whose heads spun with ideas of the sublime. Chalk cliffs drop a hundred metres and more to the sea on my left, while on my right is Alum Bay, its sands once mined for alum, now lying undisturbed in multi-coloured ridges. I am conscious that this southern coast is Britain's only margin that lies at all close to other nations. These white cliffs are pristine battlements, and scanning the sea for ships, I cannot avoid the feeling of being on sentry duty, a sense reinforced by the eruption into the downy landscape every few miles of the ruined defences of five centuries – against the Spanish, the French, the Germans.

In 1868, Thomas Huxley lectured 'On a Piece of Chalk' to the citizens of Norwich. Beginning with the chalk in his hand, he worked back through 'that long line of white cliffs to which England owes her name of Albion' to find his Darwinian theme. His disputatious claim was that

the man who should know the true history of the bit of chalk which every carpenter carries about in his breeches-pocket, though ignorant of all other history, is likely, if he will think his knowledge out to its ultimate results, to have a truer, and therefore a better, conception of this wonderful universe, and of man's relation to it, than the most learned student who is deep-read in the records of humanity and ignorant of those of Nature.

He described the microscopic skeletons of the uncountable billions of calcium carbonate algae that lived and died during the cretaceous epoch, and eventually built up from the pale silt of their decay the thick layers of England's protective chalk cliffs 'vastly older than Adam'. Huxley's geological prospect set nearby Cromer and the Garden of Eden on equal footings of chalk and clay, surely sending a frisson of pleasure through his audience. For some, though, the pleasure may have been short-lived, for all this was no more than Huxley warming to his customary theme, wielding the scientific evidence of the rocks in order to demolish the biblical version of creation.

Shakespeare seems to have sensed this cycle in which the same white mineral endlessly lives and dies. In *The Tempest*, Trinculo invites Caliban to 'put some lime upon your fingers' in preparation for their raid on Prospero's cave. But Caliban 'will have none on't. We shall lose our time, / And all be turn'd to barnacles'. It is still odd to think that lime thrown into the grave was itself once life in the form of millions of tiny marine organisms, and that our bones in their turn may become the foodstuff of future generations of shelly creatures. We may appreciate nature's cycles of water, oxygen and nitrogen but we ignore the grinding stony cycle of life-giving calcium that shifts constantly under our feet.

In his rush to scorn otherwise educated persons who unhappily remain ignorant of science, Huxley neglected to consider that quality of chalk most likely to detain the 'learned student who is deep-read in the records of humanity' – its whiteness. We tend to assume that the formal marks of human civilization are black on white, made with charcoal or graphite or the powdered carbon known as lamp black used in printer's ink. But our traces have often been the primal negative of this, urgent but judicious delineations scored on the ground in white – the finish line at the Circus Maximus, the Caucasian chalk circle that is the means of dispensing Solomonic justice in Bertolt Brecht's play of that name, the outline of a murder victim. White is there at the end, when final judgement is pronounced. In Italian, *calcio* is the word both for calcium and for the game of soccer, both meanings derived from the Latin *calx*, which is not only literal lime but also a metaphor for a goal, an achievement marked perhaps by a chalk line crossed.

Human intention lined in white is not always grimly fateful. Herman Melville in a chapter-long digression from the hunt for Moby-Dick meditates on how 'whiteness refiningly enhances beauty, as if imparting some special virtue of its own, as in marbles, japonicas, and pearls'. Two of these three, it is no surprise to find, are calcium white. Japonica is the exception: white in nature where it is not mineral – real white horses, white bears, white elephants, the albino and the albatross – is attributable not to calcium but to the arrangement of organic matter into cells in such a way that it scatters light of all colours. Melville's famous whale displays whiteness of both kinds, for while its skin is

white due to absence of other pigments, its ivory teeth are impregnated with calcium salts.

The composite structure of ivory, a tough fibrous matrix with a stony hard infill, has made it a pleasing medium for the artist. Ivory has been carved since ancient times. The seagoing Phoenicians decorated the calcareous remnants of creatures they found in and around the Mediterranean, including the tusks of hippopotami. But it was the growth of the whaling industry in the nineteenth century that gave rise to the craft of scrimshaw, an art which is romantically supposed to be the creative by-product of sailors' long hours spent on the oceans in search of leviathan. The scrimshanders' favourite medium was the massive teeth of the sperm whales that were their primary quarry, although narwhal horns and walrus tusks – both evolutionary mutations of teeth – were not wasted. They engraved images of ships and maps and patriotic subjects as well as women in mermaidenly states of undress, the fine grain of the material lending itself well to the fine-lined execution of rigging or tumbling hair, and attaining, as Melville wrote, a quality of sculpture 'as close packed in its maziness of design' as the prints of Dürer.

The most exalted material in both sculpture and architecture, together known as the monumental arts, has always been marble, the purest and whitest form of calcium carbonate that responds to the artist's chisel. Ancient Greece and Rome achieved their resplendence partly because marble quarries lay near by. Phidias used Pentelic marble from the mountains near Athens for the construction of the Parthenon, an experiment in stonework whose muscular Doric columns reflect its structural engineer's caution in adapting traditional wooden construction. Somewhat coarser in grain, Parian marble came from the island of Paros, and was employed at sites away from Attica, such as Delphi, Corinth and Cape Sounion.

Roman monuments from the Pantheon to Trajan's Column were built of marble shipped from the famous quarries at Carrara on the Tuscan coast. Carrara's Sant'Andrea cathedral is remarkable because the entire structure is made of marble – an inevitable decision perhaps, but one with the unfortunate consequence of producing an interior as grim as a cave. Other great cathedrals made more artful use of the Carrara stone, the Lego stripes of white and dark green marble that run round the exterior and interior of the thirteenth-century duomo in Siena being one striking example. My favourite Italian cathedral, however, is the one which stands like a jewelled box on the tabletop hill of Orvieto. Seen from a side street, tightly framed by ordinary houses, its west front shines with a soft white light, a glow of celestial bliss. From another angle, its Gothic finials cluster together like the sparkling skyscrapers of a great metropolis, an Emerald City, a Jerusalem indeed. Inside, the windows along the nave are not glazed but finished with thin sheets of the same marble. They admit a soothing light that casts no shadow.

Michelangelo chose Carrara marble for many of his most

important works, and made frequent excursions to Carrara to select the blocks for *David* and other sculptures from the whitest *statuario* grade marble. These visits provided a temporary refuge from whichever pope was making unreasonable demands upon him at the time. When things were going well, however, Michelangelo worked in Rome, while his favoured stone-cutter Topolino sent him consignments of stone, among which he would often include a sculptural effort of his own, to the invariable mirth of the great artist.

One project of great personal significance to Michelangelo was the tomb of Pope Julius II, begun when Julius died in 1513 and continued off and on through the reigns of five subsequent pontiffs. The work was never completed according to plan, but its several statues show the artist at his technical best. Giorgio Vasari, Michelangelo's apprentice and biographer and the sculptor of his tomb, found the figure of Moses so handsome and realistic that 'one cries out for his countenance to be veiled, so dazzling and resplendent does it appear and so perfectly has Michelangelo expressed in marble the divinity that God first infused in Moses' most holy form.'

The greatest fully realized marble creation of the Renaissance is unsurprisingly another sepulchral work: the Medici chapel

and tombs, laid out by Michelangelo and completed by Vasari. It is the prototype of modern art's 'white cube', the neutral space in which pure light reveals the truth of the artist's vision.

After Michelangelo, sculptors such as Gian Lorenzo Bernini and Antonio Canova pushed Carrara marble to new and opposing extremes of expressive excess and Classical virtue, each prizing it for the homogeneous whiteness that left the viewer with nothing to distract from the brilliance of the carving. Linked to this tradition by their choice of material, modern sculptors in marble cannot help but invoke the spirit of Classical antiquity. For Barbara Hepworth and her peers in the 1920s, determined to revive the art of stone-carving and obeisant to the dictum of 'truth to materials', marble provided the purest signal of intent. 'White was the colour of spirituality,' according to her biographer. 'In Barbara's white studio, with grey shadows, white paint, and white stone, the radio was tuned to Stravinsky and early music.' Throughout her career, Hepworth produced smooth abstract forms – single stones, pairs and triplets, stones nested or stacked, solid or pierced with holes – in alabaster, Portland stone and marble. White marble was best, always seeming to her to reflect a brighter, more Mediterranean light. Hepworth discovered the material early on when she visited Carrara and learnt to carve from a Roman *marmista*. But a trip to Greece in 1954 following the break-up of her marriage to the artist Ben Nicholson and the loss of her first son in a flying accident became a pilgrimage of artistic rededication, leading to a series of sculptures named after mythic figures and Classical sites such as Mykonos and Mycenae, executed in the most perfect translucent white marble. She selected the material to ensure the focus was always on the form, but also as a demonstration of sculpture's organic birth in the landscape, and to forge a new link in the chain that runs through Phidias and Michelangelo from the chalk figures in the hills of prehistory.

★

The cycle of life and death never stops, of course. Calcium is good for you, we are told. We are exhorted to drink milk and eat cheese in order to maintain our bones and teeth. (Chalk and cheese may be different in many respects, but they are alike in having a high calcium content.) We take calcium supplements – chalk reshaped into smooth, elongated pills like mini Hepworths or ancient sarcophagi.

Pliny tells the ultimate calcium supplement story in his *Natural History*. When Cleopatra was courting Marcus Antonius, she sought to impress the jaded Roman by announcing that she would throw the most expensive banquet ever given. The day came and the usual fare made its appearance, rich enough but hardly the ten million sesterces' worth the queen had promised. Antonius protested, and then Cleopatra summoned the main course. The servant set before her a single glass of vinegar. As Antonius grew more bemused, Cleopatra removed one of her pearl earrings, the largest pearls ever known, inherited from the kings of the East, and dropped it in the vinegar, waited for it to dissolve, drank the liquor and claimed her wager.

Literary scholars have disputed this story. Recent editions of the *Natural History* are footnoted with the received wisdom that the acetic acid of vinegar is not strong enough to dissolve pearls, and suggest that 'Cleopatra no doubt swallowed the pearl (undissolved) and subsequently recovered it in the natural course of events.' Chemists, however, disagree, and experiments using cultured pearls have shown that they will dissolve in ordinary wine vinegar to yield a potable if disgusting cocktail.

Either way, the concoction can have done no lasting harm. Cleopatra is said, of course, to have poisoned herself more effectively using an asp to take her own life when she learnt of the suicide of Marcus Antonius after his defeat at the Battle of Actium. The whereabouts of her tomb, and whether she shares it with her Roman lover, has excited great speculation among archaeologists. If found, its treasures might surpass those of Tutenkhamun and Nefertiti. The focus of attention recently has been among the limestone ruins of the temple of Isis and Osiris at Taposiris Magna south of Alexandria. The chief evidence to date is a small bust of a woman, unearthed in 2008. Unfortunately, the nose is rubbed off, making it hard to say whether or not it represents the queen of Egypt. It is carved in whitest alabaster.

## The Guild of Aerospace Welders

At his studio in rural Suffolk, David Poston greets me with a crushing handshake, and ushers me indoors. David is a jeweller and metalworker, and the reason I have come to see him is that among the chosen materials of his craft is the element titanium. The cluttered space in which I find myself looks much as you would expect a metalworking room to look. The dominant colours are grimy greys and browns. Hammers and other hand-tools lie about, and the aroma of flux pervades the air, as welcoming in its way as the smell of warm bread from a bakery.

Unusually, Poston's studio also has an upstairs, and this is laboratory-white. Under a tailored plastic dustcover in the middle of the room is his largest piece of equipment – the laser. Perhaps because they are intimidated by its reputation in aerospace and other glamorous modern industries, many craft workers regard titanium as impossible to work with. But to David, an engineer and inventor as well as a craftsman, it holds no terrors. True, it is hard, and has a higher melting point even than iron, but it has compensating virtues that make it worth the graft. It is light as well as tough, and can take on a beautiful patina.

Titanium may be cut and hammered but not soldered. Joining pieces of titanium is a matter of specialist welding, which is why David has bought the laser. He treated himself to it instead of a new car. 'Much more fun,' he says as he sits me down at the silent machine. I pass my hands through two armholes to reach into the welding chamber where I pick up two thin pieces of titanium sheet. With one in each hand, I put my eyes to the binocular viewfinder and bring them together, trying to focus on the angle they make under the crosshair sights. With trepidation, I gently depress a foot pedal to operate the laser. I feel a preparatory whoosh of argon across my fingers, sweeping away the oxygen near the metal that would cause it to burn away under the laser's heat. Then, with a sharp click-click-click, the regular pulses of the laser. An intense white flash – tinged with green unless my eyes are tricked by the bright light – bursts from the metal with each pulse. I move the metal pieces along, endeavouring to hold the angle where they meet in the crosshairs to create a reasonably tidy weld line. The temperature must reach at least 1,660 degrees Celsius to melt the metal, yet the beam is so tightly focused that I can hold the titanium pieces with my unprotected fingers just millimetres away.

★

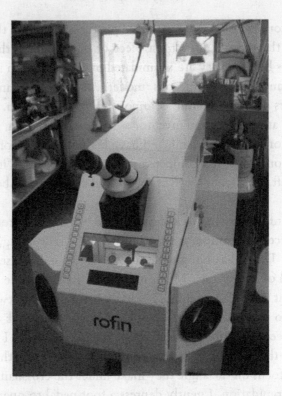

The elements with which we have the closest relationship naturally tend to be those we've known the longest. Through centuries of smelting and pouring, hammering and beating, the ancient metals have acquired more or less settled cultural associations. Gold is the universal precious metal, signifying wealth, regality and immortality. Iron is the element of manhood, strength and war. White silver is the badge of virginal purity and the feminine. Lead, tin and copper, the other metals known to the ancients, have their particular significances too. These meanings are not the product of ideal knowledge, nor of mere long acquaintance, but of man's intimate physical association over centuries of bending them to his own ends.

That it is the intimacy of the relationship that matters and not its duration is proven by the metallic elements that have been revealed by modern science. For those of them that have proven to be of the greatest utility, like zinc and aluminium, have picked up their own distinctive cultural baggage even in the relatively short time we have known them. Materials are 'culturally consequent', as the sociologist Richard Sennett has recently pointed out: 'The attribution of ethical human qualities – honesty, modesty, virtue – into materials does not aim at explanation; its purpose is to heighten our consciousness of the materials themselves and in this way to think about their value.' However abstract the human qualities attributed to various materials – grave lead, honest tin, virtuous silver – they may always be traced back to the intrinsic physical and chemical properties which the craftsman has so much time to contemplate as he struggles to shape them to his will.

What then of titanium? Despite its futuristic aura, the metal has in fact been readily available to craft workers for fifty years. Is it consolidating its meaning? 'Titanium offers lots of opportunities, but people are not being quick to find them,' David tells me. Its behaviour is well understood within heavy manufacturing industry. He describes how at Aérospatiale they weld up Airbus frames in an argon-filled hangar, the technicians going about their work wearing full breathing apparatus. This wherewithal is beyond any artist's studio, of course. But more importantly, the specialist expertise developed in these commercial environments has not been passed on in any kind of primer for general use. The secrets of the aerospace titanium welders are in effect as safe as those of the medieval guilds which once guarded the goldsmith's craft.

So people like David must rely on imagination and practical trial and error. 'It's empirical, and that's a lot of the fun,' he says gamely. Laser aside, David also uses the more traditional metalworkers' tools. He has an assortment of anvils and a kind of steel forearm that rises from one workbench like the Lady of the Lake, which he uses as a former to hammer bracelets into shape. Repeated heating and cooling gives his finished titanium pieces an instant patina, a mottled oxide coating that varies in colour from dried blood to slate and sea-green. Rigid bracelets and necklaces, their ingenious fastenings disguised within simple overall shapes, suddenly look like archaeological finds. Yet they are light – a ring feels almost weightless – and make a clatter when put down on the table, a reminder they are made of a hard new metal.

Titanium is an element in transit. It is neither so long known and so established in its various patterns of use that there has grown up around it a reactionary culture of limited expectation, nor is it so novel, scarce or otherwise esoteric that only specialists in laboratories and engineering workshops have any idea what to do with it. Though its ore was discovered in 1791, pure titanium metal was not obtained until 1910 and not made in commercial quantities until the 1950s, its potential as a strong, lightweight and corrosion-resistant metal having been demonstrated during the Second World War.

Titanium was already part of our lives – the metal used in replacement hip joints and bicycles, aeroplanes and cars, its white oxide ubiquitous in household white paint – when the Canadian architect Frank Gehry began work on the design of the Guggenheim Museum in Bilbao. Gehry explored the possibilities of the commission in his usual fashion, by making tiny models from wood and twists of paper in order to gain a quick impression of the sculptural surfaces he might use for the waterside building. Bilbao prospered during the nineteenth century through shipbuilding and steelmaking thanks to the iron ore in the surrounding Basque country, so the port city has a folk memory of huge ships blocking the vistas along its streets with walls of metal. Wishing to recapture that spirit of the place, Gehry envisioned the swooping walls of the Guggenheim covered in steel panels.

Gehry's assistants worked up his scheme using design software that had been developed for use in the aerospace industry. This computer power allowed them to reconcile the whipped-cream shapes of the building's exterior with practicalities such as the cost of the materials and the need for a sound structure. As this work progressed, somebody in the office noticed something unprecedented happening on the world's metal markets. The price of titanium dipped. Suddenly, it might be less expensive to clad the building in the exotic new metal than in conventional stainless steel. Gehry's work has always been notable for his fondness for unusual materials, and he had long admired the 'soft, buttery look' of titanium. He seized his opportunity. The completed museum, opened to rapturous acclaim in 1997, is covered in 33,000 half-millimetre-thick titanium panels – enough to clad a good-sized battleship – each one individually cut to follow the curved frame of the building. The polished surface has a tawny glow compared with the clinical coldness of steel. When steel reflects the sky it picks up its blues and greys, but titanium seems to find the warmth of the sun. The Bilbao Guggenheim has been likened to Chartres Cathedral, and certainly stands comparison with the Sydney Opera House and Frank Lloyd Wright's original Guggenheim Museum in New York, its obvious twentieth-century precursors. It has received over ten million visitors, more than delivering the boost to the regional economy that was hoped for, and jolting provincial mayors around the world into similar efforts. The building may yet prove to be Gehry's masterpiece.

How significant to the museum's success – and the city's renewal – is the metal that is responsible for everybody's first impression of the building? Its novelty, dutifully noted by every press reporter, implies bold innovation, on the part of the architect certainly, but also on the part of those who commissioned the work. The material is futuristic, and so the building becomes a monumental statement of optimism for that future. Yet the forms into which it is shaped simultaneously evoke the ship-building heritage of Bilbao, and so appear respectful of the past. Material, form and site converge to demonstrate that uncompromising modern architecture can nevertheless *belong*.

A less generous interpretation is possible, however. Marooned a couple of blocks from the life of the city, the Guggenheim Museum seems to hold itself aloof, its alien presence heightened by the wanton exoticism of its shapes and materials. It is an air-dropped package of cultural imperialism, its metal no more than a flashy façade that fails to disguise the shortage of great art within, a gratuitous flashing of a foreigner's wad of cash. The building's glittering titanium plates have been compared to fish scales, a recurrent Gehry motif. But to one critic, 'they look more like money, silver coinage pressed into the building material'.

The Walt Disney Concert Hall in Los Angeles provides an informative comparison. This, if anything, should have been the more significant building for Gehry. The project predates the Guggenheim commission – Disney money was donated and Gehry's design completed by 1991, but later delays in fund-raising and construction meant it was not finished until 2003. It was also the first major commission Gehry received from the city where he has long lived and worked; it might have been expected to represent an important milestone in the career of an architect then entering his eighth decade. Gehry had first proposed to build in stone, but the experience of the Guggenheim prompted him to change to metal cladding. No titanium here, however. The Walt Disney Concert Hall is covered in stainless steel that was found upon completion of the project to be so shiny that it had to be sanded down in order to disperse the sunbeams it was sending into nearby apartments. Critics see it as the superior work. 'The façade of Disney Hall is more refined than that of the Guggenheim, and more sumptuous, even though it is made of stainless steel, a cheaper material than titanium,' wrote Paul Goldberger in the *New Yorker*. But it has not lived up to the hype that it would surpass the Guggenheim in its global impact. There's been no Californian version of the 'Bilbao effect'. Whether it's the glow of technological optimism, or just the golden tinge of lucre, titanium clearly has something that steel does not.

## The March of the Elements

Are there elements that we consider precious or exotic today, as Parisians considered aluminium throughout most of the nineteenth century, but which will one day lose their cachet? Is titanium, for example, now on the road to banalization? And if so, what will come after it?

It seems too soon to tell where titanium will find its place. For now, it leaves too many questions unanswered. What, for instance, is titanium's gender? This question seems odd, but its answer is important if we are to know what to use it for. In culture, it has long been determined that gold and iron are masculine and silver is feminine. Titanium-branded sports gear is clearly pitched at men, but colourful anodized coatings have made the metal popular in jewellery for women. At this moment in its history at least, titanium may be masculine or feminine, both or neither. 'It liberates one from those classifications,' says David Poston.

At the Edinburgh College of Art, Ann Marie Shillito has also been using titanium for making jewellery, using its lightness and the colours it produces through anodizing to stake out an aesthetic territory some distance away from the heavier precious metals. The metal's low density (only aluminium and magnesium are lighter among the practical metals) allows her to make items such as earrings larger than they might otherwise be. Yet the fact that titanium work-hardens faster than other metals makes it very strong, too. A misplaced bend in the metal cannot simply be unbent, which makes it a demanding material to work with. Shillito has been asked to make men's wedding rings in titanium as well as earrings for women. But others are put off by the metal's space-age lightness, unable to forget the cultural conditioning that associates greater value with greater weight.

This conundrum has prompted Shillito to look again at the periodic table. 'That's when I switched to niobium,' she says. In the periodic table, niobium sits in the row below titanium, which means that it is denser. Shillito also works with tantalum, in the row below that, the row containing the real heavyweights, tungsten and gold.

Niobium and tantalum often occur together in minerals, and their discoveries were consequently a matter of some confusion and frustration, one reason why the two elements were eventually named for Tantalus, condemned by Zeus to stand under a tree whose fruit always remained out of reach, and his daughter Niobe, the goddess of tears. 'Niobium is twice the density of titanium and half that of tantalum. It is close to silver in this respect and feels more precious than titanium,' Ann Marie explains. When mass-produced titanium jewellery made it hard for her to sell her more expensive, individually made pieces in that metal, she started working wholly in niobium, which commanded a higher price because people felt it must be more valuable. But the different material also demanded a different way of working. Niobium is more forgiving than titanium, allowing Shillito to manipulate it in ribbons and sheets. Her designs in niobium appear spontaneous and free in a way that was impossible in titanium. The heavier metal also behaves more controllably during anodizing. With titanium, the artist cannot be sure what colours will be produced – Ann Marie enjoyed the element of chance that crept in after the exactitude the material demanded of her while she was shaping it. But with niobium and tantalum it is possible to tune the anodizing voltage to produce a desired colour with such accuracy that jewellery can be made to match a customer's wardrobe.

Ann Marie shows me some titanium pieces she has inlaid with niobium and tantalum. Like other precious metals, the heavier materials are relatively soft and can, using a laser, be worked like plasticine to produce decorative textured surfaces, albeit fused to a tough titanium base. The anodizing voltage produces different colours in the three metals. Into a brooch made of a sheet of titanium – brushed matt, mid grey with a hint of green – she has hammered small, bright lozenges of niobium anodized in

bright colours. Many people assume the colours are added on in some way like enamel, she says. They do not realize that they are intrinsic to the metal and its thin oxide coating – as in butterflies' wings, it is an interference effect of the light reflected from the surface that causes the colour rather than any pigment or dye. In time, perhaps, these rainbow shimmers will be seen as characteristic of these elements, just as verdigris is to copper and tarnish is to silver.

This is the march of the elements into our lives. To the Phoenicians and Romans, tin and lead were the prized new materials of the day, acquired with difficulty and danger from the remotest sources, unattended at first by any mystique or mythology, but loaded instead with the miraculous novelty of nature. Now titanium has found its way from the mines into the laboratory, and from the laboratory into the workshop and factory, and is finding its way into our culture. For niobium and tantalum, that journey is just beginning.

# Part Four: Beauty

## *Chromatic Revolution*

Clearing out some old boxes, I find my father's old Winsor & Newton artists' paints, which he used during his teenage years in the 1940s. The black metal box opens to reveal a scene of carnage. The little tin tubes of paint lie twisted like corpses in their narrow compartments, frequently stuck down with linseed oil that has split from the pigment, and occasionally caked with colour that has bled from the ruptured tubes. I turn them and read the labels: Chrome Yellow, Chrome Green, Zinc White; Terre Vert, made from iron silicate; Viridian, another chromium colour; and others too crusted or rotten to decipher. Some colours are all but banned these days, replaced by innocuous synthetic pigments not quite their equal, but in this set I find even more outré pigments such as Vermilion, the brilliant flame-red based on pure powder of the poisonous mercuric sulphide, and greens rich in arsenic.

It was one further element, however, that provided more and brighter artists' pigments than any other. Friedrich Stromeyer's discovery of cadmium was to unleash the loudest riot of colour art had ever seen, and he knew it from the first.

In 1817, Stromeyer was a professor of chemistry and pharmacy at the University of Göttingen and also held an official position as inspector of apothecaries in the state of Hanover. One of his inspections revealed that a preparation of medicinal zinc oxide was clearly not what it purported to be. When he heated the substance, Stromeyer found that it turned yellow and

then orange. This would normally indicate the presence of lead – and the need for enquiries into who was making up false remedies. But further checks were negative for lead. Stromeyer pursued his investigation to the chemical factory that had supplied the pharmacy and took a sample of the suspect material away with him for further examination at his own laboratory. Here, he deftly identified the cause of the anomaly by using a series of chemical procedures to remove the known zinc. When this was done, he was left with a pea-sized lump of a bluish-grey metal, rather like zinc in appearance, but shinier. This was the world's first glimpse of a new metal, which was duly called cadmium after the Greek for the zinc ore calamine, with which, it soon emerged, it was often found.

Stromeyer prepared sulphide of cadmium and reported that it gave a beautiful yellow colour, rich, opaque and permanent; he commended it to artists especially for its ability to mix well with blues. Cadmium was nowhere abundant, but it was reliably found in small quantities in many zinc workings, which were then growing rapidly in number to meet the demand for brassware. The sulphide soon became a commercial pigment. Its attraction was not merely convenience of supply, but the range of colours it produced – more than any other single element. According to the level of various impurities, cadmium sulphide pigments run from a slightly muddied spring green through yellow and orange to an absurdly vivid red, various deeper reds and a dark maroon – practically the entire rainbow except for blue.

These superior colours made themselves indispensable to painters. A few had quibbles about their supposed artificiality – William Holman Hunt complained that cadmium yellow 'at the best is very capricious' – but most saw the bright, pure colours for what they were. The Impressionists, Post-Impressionists and above all the Fauvists made good use of cadmium – or, it would be more accurate to say, cadmium made possible these successive waves of artistic revolution. As each new tint became available, it powered in turn the yellow sunsets of Monet, the orange-soaked Arles interiors of van Gogh and Matisse's *Red Studio*. People have romantically supposed that van Gogh was too hard up to buy the new pigments, while others believe the artist's mental state may have been affected by his use of cadmium (although he was certainly also using more noxious pigments). What is sure is that he and his peers suddenly had access to a palette of colours of an intensity never seen before.

In 1989, the United States Republican senator for Rhode Island, John Chafee, later chairman of the Senate environment committee, sought to ban the use of cadmium in pigments as part of a series of measures designed to reduce the risk of toxins from landfill waste sites leaching into ground water supplies. Sensitive souls across America found themselves torn between the interests of the environment on the one hand and artistic freedom on the other. Although the dangers of various metallic elements used in pigments were well known, the prospective legislation seemed to single out cadmium for the strongest disapprobation. One painter spoke of 'chemical censorship', and

said that having to forgo cadmium colours would be like cooking without garlic.

The vocal protests tended to obscure the fact that artists'
paints account for only a very small fraction of cadmium pigment use. Items such as colourful plastic washing bowls posed a
far greater risk if they were disposed of thoughtlessly, and for
these undemanding uses it was relatively easy to find safer pigments. But many painters felt that aesthetically there was simply
no substitute for cadmium. The sad truth of the matter was that
the wishes of artists no longer drove the pigment industry as
perhaps they did during the Renaissance, and it now seemed
that cadmium's brief reign as the painter's favourite pigment
was about to be ended.

After a lengthy campaign, though, America's artists won a
reprieve, and other countries that had been seeking to introduce
greater restrictions on cadmium soon followed suit. Today
painters are free to use their cadmium yellow and orange and
red with the same abandon as Jackson Pollock and van Gogh.
The colours – it is legally significant for their survival that they
are referred to as colours, not as paints – the colours already
bore labels in the United States advising of their chemical contents. Now they bear similar labels in European countries as
well – an improvement on the situation before when some carried the confusing message: 'HEALTH LABEL: No health labelling
required'.

There is a reason why artists were provoked to such effective
outrage, and it has nothing to do with the aesthetic merits of
cadmium colour. For it is only when a painting is *destroyed* that
the cadmium on its canvas can begin to find its way back into
the environment. Artists claim their exemption from the general ban now in force on the use of cadmium in plastics and
batteries and other mundane items based on the expectation
that their paintings will not suffer this ignominious fate. Canvas

is sufficiently expensive that artists tend to overpaint rather than throw out inferior work, and once a painting leaves the studio it tends to gain in value, which helps to ensure its survival. What had really roused America's artists, then, was not the danger to the environment posed by cadmium pigments, nor the threat that they would lose a favourite colour, but the hurtful thought that their work might not in fact be treasured for ever.

It seems beyond sad – almost a moralistic affront to our capacity for sensuous delight – that so many of the highly coloured chemicals should also be poisonous. This is true not only of the salts of cadmium but also of many long-known pigments such as yellow lead chromate and the vermilion of mercuric sulphide. Poisons in fairy tales often come in coloured bottles, or are coloured themselves. Christian Dior's counter-intuitively marketed perfume Poison exploits this mythology in a purple glass bottle shaped like an apple.

The basis for this association lies deep in evolutionary psychology and biochemistry. Humans and many species have evolved to be attracted to, yet also wary of, bright colours in nature. The colours may advertise ripe fruit and fresh meat or warn of poisonous berries and venomous creatures. Their chemical origin is generally quite different from that of man-made pigments based on heavy metals. The colours of fruit, for example, are based on yellow xanthophyll, the orange carotenes and purple anthocyanins, which are all organic compounds that contain no metallic elements. The same pigments give away the presence of such fairy-tale dangers as holly berries and the spotted red fly agaric toadstool (although the poisons they contain are not these pigments but different compounds again).

How, then, do the metal-based pigments of the artist come to be poisonous? There are various mechanisms. Some salts, such as the chromates, are powerful oxidants, which release carcinogenic oxygen radicals into the body. Others mess with biochemical pathways where vital metals such as iron and zinc are important: for example, cadmium can deprive the body of zinc by binding with certain proteins in its place; in the same way, chromium, cobalt and manganese can all displace iron from the blood plasma. The details of this biochemistry are not yet fully understood, but there is excitement that humankind might one day turn this system to its own and nature's advantage. By harnessing certain proteins, we may be able selectively to recover valuable heavy metals with which we have polluted our environment, including not only pigment elements such as cadmium and chromium but also radioactive uranium and plutonium.

Stromeyer performed his public duties well when he saved purchasers of the apothecary's contaminated zinc oxide preparation from unnecessary exposure to cadmium. Elsewhere, the danger has been revealed too late. Cadmium yellow, orange and red may be one thing, but 'cadmium blues' is quite another. This is the term that has come to describe the first cold-fever symptoms of those exposed to chronic high levels of the metal, either from soluble salts or from the inhalation of cadmium vapour. Industrial exposure poses the greatest risk. Welders working to dismantle a temporary metal structure in the unventilated space inside one of the towers of the Severn Bridge provide a grim illustration of this. The men used oxyacetylene torches to cut through bolts that were plated with cadmium. The next day they found difficulty in breathing and were taken to hospital, where one of them later died from poisoning due to inhalation of the metal vapour. In Fuchu, on the northern coast of Japan, hundreds of people succumbed to a bone-softening illness they called *itai-itai* (*itai* is Japanese for 'ouch'), which turned out to be the result of high levels of cadmium in rice that had been grown downstream of a large zinc and silver mine. Relative to these risks, the risk that cadmium poses to artists is not great: the pigments used in paints are not very soluble, and so are not very efficiently absorbed by the body even if they are ingested.

The artist's studio is not the only arena where cadmium's combination of colour and toxicity have provoked controversy. For years, I had been aware of a rumour that my local city of Norwich had received an unwelcome chemical visit in the night.

We now know this is what happened. Thursday 28 March 1963 had been a fine day, and there was almost no cloud that evening when a Devon light aircraft set out on a path that would take it from Aldeburgh on the Suffolk coast on a west-northwesterly path over the county of Norfolk. The plane was loaded with 150 pounds of a specially prepared zinc cadmium sulphide pigment, which was released at an altitude of 500 feet at the point when the plane was judged to be passing upwind of Norwich. A light southwesterly breeze dispersed the fluorescent orange particles into an invisible haze. On the ground, at forty sites in and around the city, mysterious officials – they were from the Chemical Defence Experimental Establishment at Porton Down in Wiltshire, although they bore no such insignia on their protective clothing – took up positions, ready with collectors that would allow them to count the falling particles. From declassified government papers, it appears the aim of the exercise was to test the likely efficacy of methods of biological warfare. The fluorescent cadmium pigment was merely a convenient and supposedly innocuous tracer made up in a particulate form to resemble a potential biological agent. The Ministry of Defence ran many such tests from the mid 1950s – often, so as not to arouse undue attention, over the defence establishments themselves. But sometimes the officials deemed it necessary to select a more realistic target. This was the case at Norwich, where the idea was to see whether the particles would fall to ground in an urban environment against the current of warmed air rising from the densely clustered houses. On that Thursday evening, only lowish levels of the pigment reached the collector sites. The aerial trials were repeated four times in the cold early months of 1964.

There the matter rested until news of the tests was released thirty years later, prompting fears that real dangers had been covered up. An independent report published in 2002 suggested that the risk to the public from exposure to the cadmium pigment was equivalent to what one might inhale in any city in the space of a few weeks or, less reassuringly, to smoking a hundred cigarettes, and 'should not have resulted in adverse health effects in the United Kingdom population'. A few years later, a Norwich surgeon reawakened public anxiety by suggesting that the above-average levels of oesophagal cancer that he had observed in the area might be attributable to cadmium. A Ministry of Defence spokeswoman was reported in the *Norwich Evening News* declaring in response that the trial materials were 'harmless stimulants' (an imaginative oxymoron – she presumably said, or should have said, 'simulants'). The cancer incidence was subsequently shown to be in line with what would be expected when the age and general health of the population were taken into account. In the end, the greatest actual risk may have been to the official monitors of the tests from the ultraviolet light under which they worked in order to count the fluorescent particles.

Wandering the narrow lanes of this tranquil city, with its shops devoted to music and alternative remedies, it is hard to see why it should be singled out for such an odious exercise. In fact, the ministry had first chosen Salisbury to host the trials, but it had been found too small and too hilly to produce the required thermal effect in the city air. In Norwich, I stop outside one of the numerous artists' suppliers. There, flaunting their sunflower brilliance for all to see, are tubes of cadmium sulphide paint for anyone to buy, and in doses far greater than were ever dropped from any plane upon an unsuspecting populace. All you have to do is to go in and ask.

Seeing these vivid cadmium paints forces me to consider how hard it is to describe colours at all. Our vocabulary of colour is severely limited. Red, orange, yellow, green, blue, indigo, violet doesn't begin to cover it when the average eye can discriminate several million tones. (Scientists use a particularly dodgy-sounding unit, the 'just noticeable difference', as a measure of this magnificent human capacity.) These seven colours of the rainbow tell us less about colour than they do about our laziness when it comes to naming it.

Global brands like BP and Coca-Cola cling to these primaries because it is easier to defend 'ownership' of them than subtle, in-between shades for which there is no agreed word. Beyond this there is no language of pure colour. All we can do is adopt qualifiers – light, dark, dull, greenish and so on – or seek likenesses in things that characteristically have the colour we are trying to describe. These may come from nature – primrose, say, or kingfisher blue – and sometimes they come directly from the elements themselves, as with chrome yellow or cobalt blue. But correct interpretation depends on shared cultural ground. 'Pillar-box red' is only that particular shade of red, or for that matter red at all, if you live where the pillar boxes actually are red, and where, furthermore, everybody is familiar with them. More often, the terms are hopelessly vague – sky blue, let's say – or if not, esoterically precise, such as the artists' paint called Mummy Brown, which fell rapidly out of fashion when people realized that it truly was made from ground-up Egyptian mummies.

I find myself growing more attuned to these nuances of

semantics and visual perception on a tour of the artists' paint manufacturer (or 'colourmen', as they are known in the trade) Winsor & Newton. Peter Waldron, the company's chief chemist, tells me how khaki came up in conversation one day among the many nationalities who work at the company's Harrow factory. The British staff thought they knew exactly what it meant since khaki is famously the colour of British army uniforms. I thought I did too, until I looked it up later in my dictionary and found it described as 'a light yellowish brown' – I had it down more as a muddy grey-green. The Indian workers were sure of their answer too, since *khaki* is a Hindi word meaning dust-coloured. But the French and Chinese employees were understandably more perplexed.

Such difficulties are compounded when it comes to the invention of new colours, which is an important aspect of Winsor & Newton's work. William Winsor and Henry Newton founded the business in 1832 with an innovative range of moist watercolour pigments that were easier for artists to use. The company has supplied John Constable and most British artists ever since. These days, artists' paints are a tiny part of the market for pigments, and research is limited to harnessing technology from other fields. 'We borrow from every colour-using industry – ceramics, print inks, industrial paints, food, building materials,' Peter tells me. The major effort is to replace pigments now known to be dangerously poisonous, such as those based on lead and arsenic and to some extent cadmium and chromium, with safer equivalents that artists find equal or superior to handle. 'The challenge has been to produce a range of modern colours that can reproduce anything people have done in the past.'

But artists are also interested in completely new colours. Metallic paints of the kind that have long been popular on cars are one fascination. Another wish is for ultra-bright colours that are lightfast, since most fluorescent pigments are inherently fugitive. For Winsor & Newton, it's a matter of watching the big boys and waiting for the right moment. There is some advantage in being low down the pecking order as the colourmen can at least avoid others' expensive mistakes. Peter tells me with some amusement of a bright yellow bismuth pigment enthusiastically taken up by the car industry. At first, it was not noticed how much the colour faded when exposed to the light because it faded evenly and returned to its former brightness when the light dimmed. The problem was only exposed when the test car was parked under a tree. By the time the driver returned, the finish could only be described as dappled.

## 'Lonely-chrome America'

In 1951, the Museum of Modern Art in New York put on an exhibition called 'Eight Automobiles'. Reflecting the museum's chronic fondness for European style and art, five of the eight were European designs of impeccable coachbuilding pedigree, supporting the curators' thesis that cars were – or should be – 'rolling sculpture'. The remaining three provided a representative tableau of where American design then found itself: a voluptuous 1941 Lincoln Continental, the continent in question being not America but the very same Europe, where the company president had recently spent an eye-opening vacation; a 1937 Cord 812 Sedan, which made up with chromium crusting what it lacked in fine lines; and an army Jeep as a functional alternative for those immune to the siren call of curves and glitter.

Preparation for the exhibition had begun the previous year with a conference on automobile design at which one of the curators, the architect Philip Johnson, announced – guiltily, you imagine, as if at a meeting of Alcoholics Anonymous – that he owned a brand-new Buick. Buick was the brashest of the model lines manufactured by General Motors, which also controlled the Cadillac and Chevrolet brands: 'LOOKS like a Jet Plane, TRAVELS the same way', promised one advertisement at the time. Johnson's car ran well enough, he confessed, but he was embarrassed by the gaudy look of the thing, especially when he was with his Europhile friends, who drove cars like the British MG. So in order not to offend their sensibilities or his own, he had instructed that the decorative chrome components be stripped from it.

How can one metal induce such rapture and such distaste? Though discovered as early as 1798 by Nicolas-Louis Vauquelin, chromium only became popular during the 1920s, when electroplating became widespread. Until then, nickel had been favoured for this finishing treatment. A surface layer of nickel has a gentle yellow glow, but polished chromium produces a chill blue-white colour and a piercing shine. Chrome-plated objects such as lamps and furniture were a striking feature at the influential 1925 Paris Exposition Internationale des Arts Décoratifs et Industriels Modernes, and the metal thereafter became part of the visual grammar of the Art Deco movement. It was the perfect gloss for brittle times. In *A Handful of Dust*, Evelyn Waugh's masterpiece of interwar manners, Mrs Beaver's incessant urge to redecorate other people's homes invariably involves a liberal application of chromium.

The glamorous new metal lent itself equally well to luxurious interiors and practical household objects. It provided the key signature of Art Deco extravaganzas like the Strand Palace Hotel in London. But modernist designers too made abundant use of chrome, gainsaying the puritanism so often ascribed to them. At the Bauhaus in Weimar, the artist László Moholy-Nagy brought revolution to the metal workshop, forcing its smiths to shift 'from wine jugs to lighting fixtures', abandoning craftwork in silver and gold to embrace steel, nickel and chrome-plate in designs for mass production. The skinny cruciform columns of Ludwig Mies van der Rohe's Barcelona Pavilion of 1929 – the most opulent and sensual of all temporary exhibition structures – were chrome-plated, as was much of the furniture that he designed.

The unattainable glamour of these designs, signified by their abundance of shining surfaces, merely whetted the appetite of consumers. When Parisian art deco crossed the Atlantic, to be effortlessly co-opted into the more egalitarian spirit of what America knew as the Machine Age, chrome travelled too (in style, aboard ocean liners like the *Normandie*), and was used to adorn luxury home appliances and other big-ticket items. It was not until after the Second World War that the ability to produce plate with a durable and appealing finish led to the extravagant use of chrome in many more products.

Chrome quickly became the metallic element most closely identified with the booming consumer society. It radiated modernity, glamour, excitement and speed. But it had something else too. Unlike aluminium, another material in fashion at this time, which shares some of these associations because of its lightness, chromium was seen almost always in the form of plate, and so began also to connote superficiality. For a while, though, its bright shine was enough to obliterate any doubt people might have had and give them what they craved in their lives after the Depression and the war – a little affordable lustre.

Nowhere was the consumption of chromium more conspicuous than in the automobile industry. Although the trend was global during the 1950s and 1960s, it was American cars above all that became the bejewelled emblems of the period. The man chiefly responsible for the grinning grilles and bulging bumpers and the tail fins that grew higher year by year was Harley Earl, the 'da Vinci of Detroit'. Brought into General Motors to head the corporation's newly created Art and Color Section, Earl injected Hollywood into Motown and became the acknowledged pioneer of automobile styling, exerting huge influence over General Motors' entire range of Buick, Cadillac, Pontiac and Chevrolet models. He counted Cecil B. DeMille among his acquaintances, and it soon began to show, as he introduced the concept of the 'new model year', which guaranteed his design team permanent work with its irresistible formula of change for change's sake leading to a spectacular unveiling each fall. As with the evolution of the peacock, the only path after a while was towards greater and greater excess, which meant more and more chrome. Iconoclastic gestures by puritanical museum curators were never going to be able to call a halt.

Chromium became the international calling card of American plenty. In the showstopper from Leonard Bernstein's and Stephen Sondheim's *West Side Story*, the Puerto Rican girls sing:

> Automobile in America
> Chromium steel in America
> Wire-spoke wheel in America
> Very big deal in America!

One of the girls, Rosalia, loves American things as much as the rest but pines for home, and dreams:

> I'll drive a Buick through San Juan.

Half a world away, the flash of chromium on cars is an index of the growing American presence in pre-war Shanghai in J. G. Ballard's *Empire of the Sun*. China's alliance with America is indicated by the Kuomintang pennant flying from the chromium mast of a Chrysler limousine. And when a mysterious 'Eurasian' appears near the end of the book to release Jim, the boy who is the autobiographical central character, from the stadium where he has been held with hundreds of prisoners of war, it is noted that 'He spoke with a strong but recently acquired American accent, which Jim assumed he had learned while interrogating captured American aircrews. He wore a chromium wristwatch . . .' The American impersonation is as disposable as the trophy watch.

The meanings of chromium have grown more numerous and ambiguous over time. But designers consciously employed the metal above all to convey a sense of speed – even sometimes on things that never went anywhere, like mechanical pencil sharpeners. Harley Earl's stylists streaked their Buicks and Cadillacs with fulgent fairings and elaborate horizontal corrugations guaranteed to catch the light and beam it into the eyes of admiring observers. The missile-shaped headlight mountings and stabbing ailerons were chromed too, their lines plainly intended to suggest not only speed but also an aggressive virility. These were definitely cars 'For the man of Success' in the words of one Buick advertisement of the 1950s. (The male characteristics formed in chrome find their female counterpart in the painted curves of the bodywork, making these designs into fully conceived hermaphroditic sex machines.)

The link between shiny metal and speed is apparently a permanent one. In the story of Phaëthon's chariot in Ovid's *Metamorphoses*, Phaëthon begs to borrow his father's car which he promptly crashes in flames. Its

> axle and pole were constructed of gold, and golden too
> was the rim encircling the wheels, which were fitted with spokes
> of silver.

Inaptly enough, one of the 1937 Cords was named the 'Supercharged Phaeton'.

This tendency reaches an explosive apotheosis in *Crash*, J. G. Ballard's disturbing novel in which car crashes – imagined and staged – are explored as a fetish to produce sexual arousal. Chromium serves as a stimulus throughout, providing first the prism through which erotic visions are glimpsed – 'In the chromium ashtray I saw the girl's left breast and erect nipple . . . Her sharp breasts flashed within the chromium and glass cage of the speeding car' – and then the weapon in scenes of increasingly appalling violence where the hard metal parts strike and penetrate flesh to generate sensations of sexual intensity. The harsh gleam of the metal is key. Ballard imagines 'flashing lances' of afternoon light reflected from chromium panels tearing at the skin, before moving on to 'the partial mammoplasties of elderly housewives . . . carried out by the chromium louvres of windshield assemblies' and 'the cheek of handsome youths torn on the chromium latches of quarter-lights'.

In this critique of our unreasoning love of dangerous technology, chrome is merely the surface that first excites our lust. *Crash* was published in 1973, when the first oil crisis struck and the public passion for chrome on cars was already cooling. But by this time the metal had spread its influence well beyond Paris, Weimar and Detroit to become a powerful cipher for consumerism in general.

A year or two after Philip Johnson stripped the chrome from his Buick, a group of artists and writers met at the Institute of Contemporary Art in Mayfair and resolved to take an unashamed look at the kind of thing that so offended him. The Independent Group, as it became known, counted among its founders the artists Richard Hamilton and Eduardo Paolozzi and the critic Reyner Banham. They took a more forgiving view of technology and the growing culture of consumerism, and celebrated the pulp fiction, film, advertising and mass-manufactured products that the artistic establishment chose to ignore. They sought out items possessing what they called 'symbolic content' with the idea that this, rather than patrician good taste, was the key to making things people would actually like. At one meeting, Banham explicitly praised the styling coming out of Detroit. Later, he moved to Los Angeles, where he was finally obliged to learn to drive a car, an experience he likened to acquiring Italian in order to read Dante in the original.

Richard Hamilton, one of the progenitors of Pop Art, periodically unveiled new paintings to fellow members of the group. These collage-like compositions began to incorporate the shapes of some of these shiny consumer goods. American cars were explicitly represented in works such as *Hommage à Chrysler Corp.* of 1957, whose pink and chrome mélange of sexual and machine parts simply accentuated the symbolism that was readily apparent in contemporary car advertisements. By replicating the lustre of the chrome in oil paint, Hamilton grouped himself with artists throughout the ages who have placed metal objects in their still lifes in order to demonstrate their mastery of optics and colour. But for Hamilton there was a paradox, for the greater the realism, the more the careful paint tones would imply a depth and solidity to the object. So in works where it was more important to remind the viewer of the essential superficiality of chrome plate, he made his own finishing touch superficial too, by pasting on pieces of metal foil.

With their parallels between the contours of the female body and the curves of domestic objects like toasters, these paintings might be thought to offer some kind of crypto-feminist attack on the affluent society. But Hamilton seems to be offering a more ambivalent commentary. Chrome, especially in its automotive guise, had acquired a macho appeal that persists today especially on American trucks and motorbikes. Yet, as Hamilton's next major painting was to observe, chrome was also found desirable by women. The work is called *$he* and shows in semi-abstract form part of a woman's torso, an apron, a pink fridge door hanging open, and, in the foreground, a mutant chrome appliance that appears to be part-toaster, part-vacuum cleaner. 'This relationship of woman and appliances is a fundamental theme of our culture,' Hamilton said of the work, 'as obsessive and archetypal as the Western movie gun duel.' Whatever they thought of such paintings, women were not slow to appreciate that chromium's pristine white shine represented a sharp improvement on the metals formerly used for housewares, copper and pewter, which needed frequent polishing. 'No metal, it seems to me, is quite so complete an answer to the housewife's prayer as chromium,' wrote the pre-eminent American social commentator Emily Post, who found it 'appealing not only to the eye, but to practical requirements'.

Very quickly, however, chrome seemed to change from a material promising a kind of universal glamour to one that was flash and even tawdry. Writers were first to see past the glitter. One cultural critic observed: 'there is little wrong with the American car that is not wrong with the American public', neatly inverting the nostrum of the General Motors president that 'what was good for the country was good for General Motors and vice versa'. Vladimir Nabokov describes Lolita's mother's 'depressingly bright kitchen, with its chrome glitter and Hardware and Co. Calendar and cute breakfast nook' – one of many writers' images of the territory that Don DeLillo identifies in his massive novel *Underworld* as 'lonely-chrome America'.

Chromium had lost its grip on the imagination of the socially aspirant, and the metal's reputation now fell from the cliff edge where it was perched. The fetishistic quality of polished chrome was exploited in erotic art, where the naked female body was displayed as gleaming machine. Chrome ('her pretty childface smooth as steel') is the name of a prostitute in William Gibson's short story of that name, written in 1982. Post-modernist artists like Jeff Koons gave chrome another shove on its way, recreating the kind of worthless bauble normally found dangling from a rear-view mirror on a monumental scale in polished chromium stainless steel, and relishing the irony of his super-size shiny tokens of extreme bad taste – with names like *Rabbit*, *Candy Heart* and *Balloon Dog* – selling at auction for millions of dollars. At the same time, 'chrome' surfaces themselves have grown more fake than ever as it has become possible to coat even plastics with glossy metallic finishes.

At another remove from material truth, the visual simulation of chrome – hard to achieve because the eye is acutely sensitive to irregularities in the polished surface – became a benchmark for realism in computer graphics, recorded in cult films such as *Lawnmower Man* and *Terminator 2*. Yet even the computer graphics wizards have begun to see past the surface, for, since those films were made, in the early 1990s, they have started to use 'chrome' as a term of abuse for work that strives too much for effect.

## *Abbé Suger's Sheet Sapphire*

The approach to the abbey church of St Denis outside Paris is less than promising, and the first sight of it across bleak urban plazas is little better. The building is squat, lopsided and somewhat dishevelled. But I have come for what is inside, and as soon as I have adjusted to the dimness I realize I am not to be disappointed. My first impression is of a soaring verticality, created by ranks of columns rising cleanly to the roof. Despite the charmless grey stone, the interior is light by medieval standards because of the great number of stained-glass windows and the slenderness of the piers between them. Towards the altar, a deep blue light preponderates, seeming almost to magnify the sunlight even as it transforms its colour. Other colours in the stained glass cast jewelled strands of light across the floor. The blue radiance, on the other hand, seems not so much to strike in one place as to ooze around and slowly engulf me. The effect is submarine.

St Denis is the prototype of the Gothic cathedral, the magnificent creation of the famous Abbot Suger. We tend to think of Gothic architecture as heavy and spooky, but this is not the case here. The blue glass, one of many beautiful and novel materials that Suger employed, is concentrated for maximum effect in the windows at the east end of the cathedral, where the expectant gaze of worshippers is answered by the morning sun. Suger said the church 'shone with a marvellous, uninterrupted light'.

Where some of the windows were restored in the nineteenth century, the colours are brighter in the replaced panes, and the detail is sharper where they have been etched. But the authentic Gothic blue remains quite as intense as the new. It is clear from the nativity window that its medieval craftsmen knew the colour was special: Christ himself is swaddled in the rich blue, and Mary, too, is draped in it.

Blue has always been one of the hardest colours to extract from nature, often seeming as intangible as the sky itself. But Suger was able to take advantage of newfound sources of topquality blue, which was obtained from ores of the as yet unknown metal cobalt. Cobalt compounds can attain an intensity of colour five times that of any other colorant of glass, and the availability of these exceptional minerals sparked a remarkable fashion for blue in the twelfth century. Following the example of St Denis, first Chartres, then Le Mans and other great churches of the period flaunted 'precious sheets of sapphire' in their windows. Inspired by the glass-makers, other crafts began to make more frequent use of blue in enamelwork, painting, clothing and heraldry. The colour came to be favoured for the Virgin Mary's dress, and through this holy association was also adopted by the French monarchy. I realize as I leave St

Denis and make my way back into Paris that this blue is all over the city: on the traditional blue-and-white enamel street names and on the signs in the Métro.

By the end of the century, demand for blue glass was so great that other blues, derived from copper and manganese, had to be employed to meet ecclesiastical demand. But while these less stable tones have deteriorated over the centuries, the cobalt blue at St Denis and wherever else it has been used has remained as true and intense as in Suger's day, its 'luminous darkness' considered by some to be the perfect representation of 'Divine presence'.

By analysing its characteristic impurities, it is possible in principle to trace any mineral back to its source, like a detective analysing soil from a shoe. In practice, though, the work of matching the elements found in finished artefacts with the composition of specific mine ores has hardly begun. It seems likely, however, that Suger's blue came one way or another from mines in Persia. Traders may have carried raw smaltite ore – or the glassy derivative of it known as smalt – directly to France, but it is impossible to be sure of this as medieval glass was often made from recycled Roman glass and Byzantine mosaic tiles, whose raw materials would have come from the same Persian sources.

Smaltite is a shiny grey mineral that offers little clue to the intense colour that lies hidden within it. The cobalt oxide that is obtained by roasting it in air is also dull in appearance. Only when this material is fused together with quartz or potash does it form the bright blue smalt. As a glassy material itself, smalt is perfect for fusing into glass and ceramics, but despite its intense colour it is less suitable as a pigment in paint. If ground too fine, it begins to scatter all light rather than just reflecting the blue, and this makes it appear pale. But too coarse-ground, it leads to a streaky finish in oil paint. Nevertheless, smalt pigments were used by sixteenth-century artists often as a base or thinly dispersed in painted skies. Painters such as Titian, who conspicuously included many blue garments in their paintings and who did use smalt, still preferred to use ultramarine, made from lapis lazuli, for the finish.

I buy a small pot of smalt from an artists' supplier. It is not a powder like other pigments, but grainy to the touch like a very fine sand. Under a bright light, I can see that the intense blue is subtly modulated: the colour of the material itself is darker than I first perceived but is lightened by the sparkle from its crystal grains. I mix some with linseed oil, working like the artists of the Renaissance. The mixture crunches gently under my spatula and darkens almost to black as the liquid spreads through the pigment. Colour returns as I spread the mixed paint on canvas, but no matter how thinly I do this I cannot produce a pale blue, only ever scratchier flecks and streaks of the intense original colour.

A new, European, source advanced the popularity of blue in the sixteenth century, when it was found that the long-established silver mines in the mountains between Saxony and Bohemia were also rich in smaltite. The Saxon ore miners, traditionally regarded as the best in Europe, hated the toil of extracting the new mineral, however. The labour was hard and involved exposure to harmful fumes, released when the ore's other main ingredient, arsenic, was roasted off. The miners blamed their woes on a little earth demon named Kobold.

When Goethe's *Faust* first summons the figure of Mephistopheles, he invokes 'the elemental four' of fire, air, water and earth in turn, with earth personified by this evil spirit:

> First, to confront this thing of hell,
> I must repeat the four-fold spell:
> Salamander bright shall burn,
> Sylph invisible shall turn,
> Undine flow within her wave,
> Kobold shall slave.

The Norwegian composer Edvard Grieg titled a growling, aggressive little piano scherzo of his 'Smatrold' (meaning 'little troll'); the Germans call the same piece 'Kobold', but the chosen English translation, 'Puck', fails to capture the real nastiness of the character.

Of course, the fact that its chemical constitution wasn't properly known and that the element behind it lacked a name of its own did nothing to prevent the blue pigment from becoming a highly marketable commodity. Cobalt prospered anonymously for centuries before it was eventually discovered, in around 1735, when the chemist and controller of the Swedish mint Georg Brandt divined that smaltite was not merely the compound of known metals and arsenic that had been assumed. He named the new metal cobalt after this incubus of the underground in tribute to those unlucky miners – and perhaps also as a way of wresting this name from its pagan associations and fastening it instead to the shield of Enlightenment science.

Smalt was highly compatible not only with glass-making but also with the materials and processes of pottery. It was one of the few substances that kept its colour when the pottery was fired. Indeed, heat intensified the blue. Other colours could always be painted on afterwards, but the opportunity to seal in the colour beneath the glaze guaranteed its dominance. The earliest European wares to use this blue in their designs, such as majolica and faience, relied on cobalt from Persia, as did the glass-makers of Venice. Though compositions vary, the blue glasses and ceramics used in Islamic and Christian decorative arts of the medieval period also contain cobalt from this fruitful source. The jewel of Persian civilization is the public mosque that Shah Abbas I built at the beginning of the seventeenth century in Isfahan, its façade a blaze of golden Arabic script set against this same blue glaze. The Chinese, who had been creating blue designs in their ceramics since the ninth century, also relied on Persian supplies of 'Mohammedan blue', transported along the Silk Road. This art reached its apogee during the Ming Dynasty. The craftsmen of this period often restricted their palette to this single colour, preferring its stylized contrast with the icy white of the porcelain to more naturally coloured scenes.

In my mind's eye I have an image of dusty blue tracks radiating east and west from the mines of Persia and Saxony to the great artistic centres of the world, like the starbursts on an airline route map. The lines kept on spreading. European potters, inspired by the Ming porcelain first introduced to Portugal by returning explorers and then imported in greater quantities by the Dutch East India Company, sought to emulate the Chinese artistry using the local smalt from Saxony. Delft became the centre of this activity, and the city's name has become synonymous with the blue-and-white pottery that was created throughout the Netherlands at this time. In 1708, a new mix of clays and hotter kilns finally enabled Europe to rival Chinese porcelain. Royal Saxon porcelain produced at Meissen, close to sources of smalt as well as the new clays, soon came in many patterns, but one of the most enduring is Blue Onion, a floral design loosely based on a Chinese original. Other porcelain factories quickly sprang up throughout Europe. Despite the advent of pigments in other colours, the blue designs that many of them have produced from the outset remain some of their most popular.

In Britain, too, Chinese-inspired blue designs were an early success for the eighteenth-century porcelain factories of Royal Worcester and Spode, which found ways to industrialize the production of tableware. One of these designs, Willow Pattern, is still sold today. The first English maker of fine porcelain was William Cookworthy, a Plymouth pharmacist who founded his own works in order to exploit the china clay deposits he had discovered in Cornwall. He was also to be the key to the smalt trade in Britain. Cookworthy's trade and location were essential to his achievement. In the busy naval port, he had the pick of exotic raw materials from Britain's overseas colonies and foreign lands. He bought pottery clay from Virginia and chemicals to make up the medicines that he then supplied to the navy and other outbound shipping. He recognized the superiority of Saxony smalt and moved to gain a monopoly of the material imported to Britain. When Cookworthy later shifted his own china production to Bristol, he was in a good position to control the supply of imported smalt taken up the River Severn into the heart of the Potteries, where, in 1784, the manufacturer Spode began transfer-printing images using a cobalt underglaze to create the most distinctive of all English blue tableware.

Cookworthy's smalt soon found favour with the city's glass industry, too. Bristol was one of the vertices in the 'triangular trade' linking Britain with Africa and the Caribbean via the slave trade. The sugar arriving at Bristol from plantations in Britain's Caribbean colonies provided an incentive to set up local distilleries, and these in turn generated a demand for bottles. Bottles were one of the many manufactured goods then exported to Africa and elsewhere, completing the infamous triangle.

Although coloured glass and clear glass are similar in chemical composition and basic properties, coloured glass attracted a lower excise duty than clear glass, which was usually destined for the table, or for windows and chandeliers. In order to avoid the higher duty, bottle-glass-makers made sure to colour their glass. By the second half of the eighteenth century Bristol was famed for its coloured glasses: as well as greens and browns due to iron impurities, they developed a deep blue glass from Cookworthy's smalt. Most bottle glasses didn't merit a second glance, but the blue glass, which was novel and unquestionably beautiful, quickly found a market. This glass was made into decanters and stemware in the height of Georgian style, appealing to the prosperous merchants of the city and the *nouveaux riches* of nearby Bath and beyond. The boom was short-lived, however. The loss of the American colonies after the War of Independence precipitated the collapse of the entire industry, leaving behind little more than the phrase Bristol Blue.

In 1996, Harvey's, a sherry importer long based in the city (now swallowed up by some placeless international conglomerate), decided to commemorate its centenary by putting its most popular Bristol Cream sherry in blue bottles. The marketing idea was a great success, and the sherry comes arrayed in Bristol Blue to this day.

I bought my smalt from a famous old artists' materials shop, J. Cornelissen and Son in Bloomsbury. The place looks much as it must have done when Monet and Pissarro bought supplies for their London townscapes and André Derain selected the flamboyant new cadmium colours for his psychedelic vision, *The Pool of London*. Black wooden shelves rise on all sides from the plain board floor right to the ceiling. Tubes of paint are set out in racks at eye height, but the most striking feature is the rows of huge glass-stoppered bottles above them, containing violent shades of powdered pigment. This is colour as pure as it comes. The cobalt blue was brighter and paler than my sandy smalt, with a distinct hint of red. Alongside it was manganese blue, a fabulously bright, green-tainted blue based on barium manganate, as well as chrome yellows and greens, a wide range of brilliant cadmiums and cobalt violet, a confectionery shade so improbable that it could hardly be imagined to have a natural origin.

Later I pay a visit to Cornelissen's warehouse to see the pig-

ment buyer. A curious affinity must have drawn him to work here, for his name is Ole Corneliussen. Danish by birth, he insists he is unrelated to the Belgian who established the business in 1855, and he spells his name differently. I am mildly disappointed to learn that his job does not involve touring mineral deposits in far-off lands. 'I don't know if you've seen the spice market in Istanbul,' Ole ventures a little ruefully. I nod. 'It's not like that.' Instead, samples are requested from the manufacturer and sent in for approval; where the pigment was extracted or refined seldom figures in the buying decision. Siennas may still come from near Siena and copper colours such Terre Vert from Cyprus, but material quality always overrides sentimental history.

All the old pigments are still obtainable, even orpiment and realgar, the ancient yellows and reds based on deeply unpleasant sulphides of arsenic, which are bought by specialists restoring old artworks. Often the colours are not quite as they first appear to be. Even black and white are not black-and-white. Ordinary lamp black, the carbon powder traditionally made by burning oil lamps, I see, is not really black but a very deep blue-grey; spinel black, based on manganese and copper oxides, is far blacker. Ole shows me some of the last Flake White he will be able to sell before new European health and safety regulations come into force, after which artists will have to make do with titanium white. They are not all happy at the prospect. 'Titanium is very sticky when you grind it, whereas Flake White has this elastic feel you get from the lead,' he explains. Flake White is lead carbonate made from a mille-feuille of sheet lead and chalk layers – hence 'flake'. The mixed paint feels heavy on the brush because it is so dense, and handles and dries in a way that artists love.

Ole Corneliussen himself doesn't paint, so his appreciation is purely for the pigments – 'mainly for the colour, but it's not always very easy to describe' – and for the occasional thrill of tracking down some rare oddity. One of his favourites is the sugary cobalt violet I had noticed in the shop. It is both bright and intense, 'one of very few colours to have that quality, it's a very deep violet even though it is light in tone – very hard to describe without using the word fluorescent'.

I leave him to attend to an order from the artist Anish Kapoor for a tonne of calcium carbonate whiting – 'Goodness knows what he's making.'

### Inheritance Powder

Arsenic, wrote Gustave Flaubert in his *Dictionary of Received Ideas*, 'Can be found anywhere (remember Mme Lafarge). There are certain populations that eat it regularly!'

As usual, Flaubert, the surgeon's son, is sharp on scientific matters. Arsenic is widespread and abundant – so much so that it is never mined for itself but is obtained in plenty from the waste of other mining activity – and for all its deserved reputation as a poison it is essential to human biology. It is not only eaten, particularly in shellfish, but has a long and distinguished history as a medicine that continues to this day. By the nineteenth century, arsenic compounds were employed as pigments and dyes, in many medicinal preparations, alloyed with lead in shot, and in glass-making and fireworks.

But it is as the classic poison agent that arsenic is best known, and of the many tales of poisoning by arsenic, both imagined and true, that which surrounds the death of Napoleon on the remote South Atlantic island of St Helena is surely the most contentious. The story shows once again how colour and toxicity are bound together in nature. When the deposed emperor died in May 1821, the autopsy conducted by his personal physician, one of the retinue who had accompanied him into exile and a fellow Corsican, found a stomach ulcer and gave the cause of death as stomach cancer. It was not until the diary of the emperor's valet was published much later, in 1955, that doubts began to be raised. To Ben Weider, a Canadian Napoleon enthusiast, the descriptions it gave of the emperor's deteriorating condition in the early months of 1821 seemed remarkably like the symptoms of poisoning. In 1961, Sten Forshufvud, a Swedish toxicologist, performed analytical tests on samples of hair – many of Napoleon's loyal servants had been prescient enough to snip off an imperial lock – and found they did indeed contain high levels of arsenic. The two men eventually teamed up and conducted further tests in pursuit of the theory that Napoleon had been a victim of deliberate poisoning, and, through a tortuous series of murder-mystery deductions, came to a definite conclusion as to whodunit. Neglecting to ask themselves too many more questions, Weider and Forshufvud amplified their theory in a series of books.

The ensuing publicity led the chemist David Jones – the author of the Daedalus column of conceivable scientific fancies in *New Scientist* – to wonder whether in fact the wallpaper at Longwood House, where Napoleon had been kept prisoner on St Helena, might not be a more plausible source of the poisonous arsenic than an assassin. Green shades in wallpaper of the time were frequently made by using compounds of arsenic, following the discovery by Carl Scheele of copper arsenite, a colour that became known as Scheele's Green. By the time of Napoleon's exile, a bright new green was also available, based on the aceto-arsenite of copper – the lucky product of colourmen's natural urge to see what would happen if you combined copper acetate, the long-used pigment known as verdigris, with Scheele's murkier tone.

This colour is so striking that it was marketed under the name Emerald Green. Because of its poisonous properties, it is no longer sold, but I find a small tube of it among my father's paints, its label translucent from sixty years' absorption of linseed oil. To my surprise, the knurled metal cap yields immediately and the paint inside glistens willingly. The colour is lurid and has a bluish grey undertone, which marks it out as quite alien from any shade in nature. This eye-aching, sickly green makes me wonder if, when we refer to a 'poisonous shade', it is not these arsenic pigments that are responsible for the phrase.

Jones knew that under the right conditions the arsenic in such materials can be chemically converted into gaseous forms such as the hydride arsine. Speaking by chance on a radio programme about this phenomenon, and how it might account for many mysterious illnesses and deaths throughout the nineteenth century, he speculated that perhaps Napoleon's death on his humid prison island was also hastened in this way. If only the colour of the Longwood wallpaper were known, it would help to establish the facts. To Jones's great surprise, he received a letter in response to the broadcast from a woman who not only knew the colour of the paper, but had a sample of it in a scrapbook recording the travels of a family ancestor. In the scrapbook was a page of souvenirs from a visit to St Helena in 1823, among them 'a piece of paper taken off the room in which the spirit of Napoleon returned to God who gave it'. Jones published the results of a chemical analysis of the green-and-gold star-patterned paper in 1982 in *Nature*, confirming the presence of arsenic – a not unexpected result given the popularity of the colour at the time. At the same time, doubts were thrown on to Forshufvud's original analysis. New tests on the emperor's hair using more sophisticated equipment showed high levels of antimony and other potentially harmful elements as well as arsenic. The antimony probably came from a standard emetic administered to Napoleon, and it is quite likely that the medicine did him more harm than good.

Nearly 200 years later, it is impossible to establish cause and effect with any certainty, and even the basic precaution of a DNA test to authenticate the sampled hair remains to be done. Nevertheless, recent biographies of Napoleon concede that his symptoms were consistent with arsenic poisoning, and that arsenic from whatever source may have been one factor in the former emperor's death. The consensus is that there probably was some attempt on the part of his British custodians to disguise the real cause of death as part of a wider cover-up – their mis-management of the island had allowed dysentery to run rife – but that there is no need for wild assassination theories.

The most recent re-examination of the evidence in 2008 found that Napoleon's hair from periods of his life predating his exile, as well as the hair of his wife, Josephine, and other family members, all had levels of arsenic that would be regarded as ele-vated by today's standards. There was no evidence of a sudden rise in the arsenic concentration following his incarceration, as would have been produced by deliberate poisoning. However, rather than putting themselves to the trouble of rounding up Napoleon's locks for analysis, the authors of this latest study might simply have surveyed the toxicological literature. They would have found that human remains of this period in general can exhibit levels of arsenic that would be classified as dangerous by today's standards, reflecting nothing more than the fact that the element then was indeed 'found anywhere'.

Arsenic may or may not have contributed to the death of Napo-leon, but it has certainly been the chemical agent responsible for many other cases of poisoning, both deliberate and accidental. The one that most closely follows the Longwood wallpaper scen-ario concerns Clare Boothe Luce, the United States ambassador to Italy in the 1950s, who was slowly poisoned – accidentally, it was later established – by flakes of paint falling from the ornate ceilings of the ambassadorial residence. She retired due to illness and made a recovery later. Luce was an unlucky latterday victim of a widespread danger. Green paint, colour printing and col-oured papers, green in wallpaper, green dyed furnishings and clothes, and especially the green colouring used for the foliage in artificial flowers all contained compounds of arsenic and probably accounted for many mysterious deaths in damp bed-rooms and nurseries. There were rising suspicions during the

Victorian period that these materials were to blame. The *Lancet* and the *British Medical Journal* sounded the alarm and campaigned vigorously against arsenic, but while a few companies began to advertise arsenic-free papers, the decorating industry for the most part decried the idea that their products could release any harmful substance at ordinary room temperatures. It was not until 1893 that it was shown that arsine gas could be generated by the reaction of mould from wallpaper paste with the green colorant. In an essay on the art of dyeing in that year, the designer William Morris railed against synthetic dyes – among which were the arsenic greens – for 'doing great service to capitalists in their hunt after profits' but leaving the domestic craft 'terribly injured' and 'nearly destroyed'. Morris fought noisily for the survival of traditional vegetable dyes in wallpapers and textiles. Odd, then, to find that recent X-ray analysis of Morris's own wallpaper designs has revealed that his green came from copper arsenite, while a red rose in the pattern was the vermilion of mercuric sulphide – 'a very dangerous piece of art!'

Others took arsenic knowing full well what they were doing. Romantic before his time, the teenage poet Thomas Chatterton

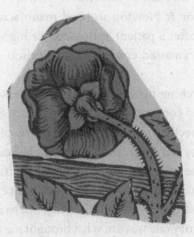

used arsenic to commit suicide in 1770. At Tulle in the Limousin region of France, Marie Lafarge was tried and found guilty of poisoning her husband with arsenic in 1840. The case was such a cause célèbre that Flaubert could safely include it in his dictionary more than thirty years later, knowing that his readers would remember the episode. The author had more than a passing interest in desperate housewives, of course: his own creation, Emma Bovary, also uses arsenic to commit suicide. Madame Lafarge was convicted when the evidence of the brilliant toxicologist Mathieu Orfila, called by the lawyer in her *defence*, showed there to be arsenic in her husband's exhumed body and in food residues. It was the first case in which forensic chemistry was used to secure a verdict.

Both in fact and in fiction, where it became a staple of detective stories, arsenic was generally obtained from pharmacies, where it was widely sold as everything from medicine to rat poison. The form of the element used in these cases is likely to have been the sugar-like oxide known as white arsenic, a colourless substance of no interest to the decorative arts. This became so well known for its use in family murder cases that it soon acquired the nickname 'inheritance powder'. As for Emerald Green, Winsor & Newton stopped manufacturing the colour around 1970 after a patient at Broadmoor high-security psychiatric hospital amassed enough of it in prison art classes to kill himself.

While searching for deaths attributed to arsenic poisoning, I was startled to come across the story of Mary Stannard of New Haven, Connecticut. In 1878, she was murdered in her twenty-second year by her lover, the Reverend Herbert Hayden, when it appeared she might be pregnant. He administered a large dose of what she took to be a treatment to induce an abortion but was in fact arsenic. He then bludgeoned her to death and slit her throat. This gory tale was not what brought me to a halt, though.

No, what stopped me was that Mary and Stannard are the two forenames of my own mother, Connecticut-born in 1930. Was this a branch of my own family tree that had been so brutally severed?

Before the twentieth century, public access to arsenic was largely unrestricted. Today, white arsenic is more closely guarded, but still widely used in medicine: the United States Food and Drug Administration recently approved it for use in treating patients with leukaemia. Arsenic in nature is less readily corralled, and here its compounds quietly do great harm. The drinking water of up to 100 million people around the world may be tainted with it. Surveys of the water, soils and rice grain of Bangladesh have shown levels of the element far above the limits judged safe in the West, themselves rather arbitrarily set in response to public outrage at the wallpaper deaths. The phenomenon is recent and has been traced to the changeover from deep-water wells to so-called tubewells driven into shallow river sediments. These produce potable water for millions of people, but the water contains arsenic washed out of natural deposits upstream. Some scientists believe that a cancer epidemic is now inevitable as a result. It is not what Flaubert had in mind, but it is unfortunately true, and on a far larger scale than he imagined, that there are populations who eat it regularly.

## Rainbows in the Blood

Slipped in among Lee Chong's grocery and the Palace Flophouse, one of the varied emporia of Monterey that John Steinbeck describes in *Cannery Row* is the laboratory of Western Biological, where you could buy 'the lovely animals of the sea, the sponges, tunicates, anemones, the stars and buttlestars, and sun stars, the bivalves, barnacles, the worms and shells, the fabulous and multiform little brothers, the living moving flowers of the sea', and much more.

Specimen collectors have always marvelled at forms of life from under the sea, which are so often beautiful and puzzling, uncertainly poised at the junction of the animal, vegetable and mineral worlds, and only delivered up from the depths at irregular intervals by stranding storms. The most mysterious items on Steinbeck's list are the tunicates, a class of animal that includes the sea squirts, which normally live on the seabed in colourful clusters of bag-like organisms. I once arranged to borrow a tunicate specimen from the Natural History Museum for an exhibition. It came in a squarish tank of thick glass filled with preserving liquor like the specimens at Western Biological. The creature, or creatures, or growth – scientists still aren't really sure how to classify the things – was a chaotic eruption of shapes and colours like some absurd dinner-table centrepiece. Each 'bag' wears its own transparent tunic like a plastic mac as it puffs gently in and out, pumping seawater in order to extract nutrients. The organisms are dependent for some biological functions on the cluster as a whole, but manage nevertheless to express their individuality in blue, green, purple, pink, yellow and white.

In 1911, a German physiologist named Martin Henze, who was curious to learn why they adopt these seemingly indiscriminate hues, drew some tunicates from the Bay of Naples and was surprised to discover extraordinary quantities of the element vanadium in their blood. Positioned one place before chromium in the periodic table, vanadium, like chromium, forms compounds that exhibit a wide range of colours. Vanadium in these creatures can be a hundred times more concentrated than in the seawater from which they suck their food, and, according to scientists at the University of Hiroshima, tunicates may possess the highest concentrating ability for any metal of any animal. It seems reasonable to assume the vanadium is harvested for some purpose, but despite pinpointing the green cells, known as vanadocytes, where the element collects in the blood and identifying various proteins that bind to it, the scientists are still unsure what that purpose is. At first it was thought that the vanadium might have a function analogous to the iron in our own blood, but this notion has been discounted; it is possible that the element plays a part in the animals' immune system.

This bizarre anomaly of nature came to the attention of military officials during the Second World War. Vanadium produces a much tougher steel than other metals and was therefore in demand for use in soldiers' helmets and armour plating as well as in machinery. The United States War Department approached Donald Abbott of Hopkins Marine Station – the research outpost of Stanford University at Monterey that Steinbeck used as the model for Western Biological – wanting to know whether tunicates might be gathered or even farmed for the exotic metal. The government men flattered the scientist that the vanadium was needed not for conventional armour, but for the top-secret atomic bomb project. Abbott presumably set to work on the problem, but nothing more was said about it. Asked about the episode many years later, Abbott's widow, Isabella, also a scientist at the station, confirmed in the obscure technical bulletin *Ascidian News*: 'Such a request was made of Don, but he showed them how much vanadium was in the tunicates that took it up, and it was just too small to bother with, and as I remember that was the end of it.' But perhaps vanadium was not the real goal. During the war, 'vanadium mining' was the code term used to describe the search for uranium ores needed for the atomic bomb. (The two elements occur together in some minerals, a fact noted in the name of Uravan in western Colorado, one of the mining sites where this subterfuge was in operation.) It may be that the War Department wondered whether the tunicates might be used to concentrate uranium too.

Vanadium was discovered twice and was named on both occasions in homage to its colourful chemistry. In 1801, only three years after Nicolas-Louis Vauquelin had discovered chromium in Paris, Andrés Manuel del Río, a Spanish-born mineralogist at the School of Mines in Mexico City, identified the new element in one of the many unfamiliar minerals that came into his laboratory. Delighted by the many colours of its salts, he named it panchromium. A couple of years later, the explorer and naturalist Alexander von Humboldt visited Mexico and took back samples of the mineral to be tested in Paris. One of Vauquelin's colleagues analysed the substance and declared that it was nothing more than chromium. Del Río bowed to this judgement, unaware for many years afterwards that the French science was flawed, and that documents he had sent separately, which would have provided stronger support for his claim to discovery, had been lost in a shipwreck.

It was not until 1831 that the element was rediscovered half a world away in a quite different kind of mineral by the Swede Nils Sefström and given the name by which we know it today. Sefström was the director of mines at Falun, 200 kilometres

north-west of Stockholm. He had formerly worked as an assistant to Jöns Jacob Berzelius, one of the greatest figures in the
history of science, who, as we shall see later, played his own
disproportionate role in the discovery of the elements. It was
Berzelius who chose the name vanadium, after Vanadis, an
alternative name for the goddess Freyja, who appears in some of
the Norse *eddas*. Vanadis (the *dis* of the Vanir, or 'lady of the
beautiful people') is the goddess of love, beauty and fertility.
Except when engaged in some naked seduction, which is often,
she appears robed in colour and glittering with jewels. Her prize
possession is Brisingamen, the necklace of the Brisings, which is
represented with the most elaborate gold craftsmanship and frequently studded with flaming gemstones. When she weeps, her
tears are of red gold if they fall on solid ground and amber if
they fall at sea.

The vanadium mineral – an ore with an unpredictable yield
of iron that sometimes proved strong but sometimes brittle –
had been a puzzle to Berzelius for some time. In 1823, it was
examined by the German Friedrich Wöhler, the most famous of
the many chemists who beat a path to Berzelius's laboratory.
Wöhler later became the first person to synthesize a substance
found in living organisms (urea, a simple end product of protein
decomposition) from exclusively mineral precursors, thereby
proving that chemistry was universal across the animate and
inanimate realms. But on this occasion there was no revelation.
When Sefström duly made his breakthrough, Berzelius wrote to
Wöhler with his own little prose *edda*:

Long ago there lived in the far North the goddess Vanadis, beautiful and
alluring. One day there came a knock at her door. The goddess sat quietly
and thought, 'I will let him knock once more', but the second knock
failed to come, and the man who had knocked merely walked away. The
goddess was curious to know who was so indifferent to being admitted,

and she sprang to the window to view the departing guest. 'Ah-ha!', she said to herself, 'it's that rogue Wöhler. It serves him right; had he been a little more persistent I would have let him in. But he doesn't even look up at the window in passing.' A few days later there was another knock at the door. Sefström stepped in, and from this meeting Vanadium was born.

The name of an element can confer a kind of immortality. For a start, the unfortunate Del Río might be better known today if a rival proposal to name his discovery rionium had won more support. But even deities stand to gain by chemical association. 'In his naming of the elements Berzelius gave new life to the figures of Scandinavian mythology,' according to one of his biographers. 'Thorium and vanadium will remain in the periodic table long after Thor and Vanadis and the other gods and goddesses of the Vikings have been forgotten.'

Preserved in the collection of the Berzelius Museum in Stockholm are some three dozen test tubes filled with the various vanadium salts that the Swede had been able to make. The colours include bright turquoise and pale sky blue, orange, maroon, chestnut and tan, various ochres, a sludgy green and black – many of the shades found in the tunicates.

## Crushing Emeralds

Beauty comes out of necessity. For though we may dress up the truth with fancy aesthetic theories, we are biologically programmed to appreciate colour and the reflected glare of the sun for our survival. These things are signals of ripe fruit in the trees and the sparkle of fresh water. No wonder that Vanadis named her daughters – with wince-making new-age trendiness, it seems to us now – Hnoss (Jewel) and Gersemi (Treasure), reflecting these two properties so coveted in the dull, dark north: the colourful and the gleaming.

Prized above all are finds or artefacts that unite the two qualities, such as polished gemstones and of course the shining yellow metal gold. These conjoint desires are reflected in our language. The word gleam stems from an Indo-European root, *ghlei-*, *ghlo-* or *ghel-*, meaning 'to shine, glitter or glow', which is also the origin of the word yellow. An astonishing number of words that describe light coming in bright flashes share this root (glint, glitter, glimmer, glisten, glitz, glance and gloss among them, as well as cognates such as glad and gloat, which reveal our emotional investment in objects possessing this property). Glass as well as glare comes from the Anglo-Saxon *glær*, meaning amber, another shining yellow substance found in nature and one of the customary ornaments of the goddess Vanadis.

The Viking goldsmith unites metallic gleam and crystal colour when he sets a stone – Brisingamen is described as 'gem-figured filigree' in *Beowulf*. But what the smith cannot know is that both metal and jewel may have the same elemental origin. Vauquelin had discovered bright chromium by accident in a humble if rare specimen of red lead carbonate from Siberia. Along with other scientists of the age he was greatly preoccupied with the question of what gave precious stones their signal colours. In the vast chemical encyclopedia that he produced with his mentor Antoine-François de Fourcroy between 1786 and 1815, Vauquelin agreed that the ruby was 'the most esteemed of precious stones', and noted that the beryls, a class of gemstone that he recognized to include emeralds, came in colours ranging all the way from blue-green to the 'russet yellow of honey'; 'the best emeralds come from Peru,' he added.

Shortly after his discovery of chromium, Vauquelin, newly promoted as an official assayer of precious metals, was to be found pounding a Peruvian emerald in a pestle and mortar and dissolving its powder in nitric acid in an attempt to unweave the rainbow of the jewel box. He was able to convert the residue into the same substance he had obtained from the Siberian ore, thereby proving that the colouring agent in emerald was chromium. He went on to show that the red of ruby was also due to chromium. More comprehensive analysis only possible more than a century later finally explained why these gems have been prized for so long. The deep red of rubies and limpid green of emeralds is only the half of it: the chromium in both stones also fluoresces with red light, so that the stones appear to flicker with inner fire.

If the same contaminating metal, chromium, could be responsible for two such brilliantly contrasting colours, it suggested that there was something worth investigating about the basic matrix of the ruby and beryl crystals into which the chromium was locked which might explain this dramatic difference. Vauquelin returned to analyse the beryls in more detail, discovering that they were comprised of a number of basic ores. The main constituent was silica, or silicon dioxide, as in sand, quartz and amethyst. Alumina made up much of the remainder. This crystalline form of aluminium oxide is the principal ingredient of corundum, of which rubies and sapphires are made. But there was also, Vauquelin now realized, a new oxide which had escaped detection earlier because of its unremarkable similarity to the others. Isolated and purified, however, this oxide did possess one exceptional property. It was sweet to the taste, and for this reason Vauquelin named it 'glucina'. The new metallic element that he knew it must contain he called 'glucinum', although nobody would be able to produce it for another thirty years. (Zirconium, another new element, discovered in rather similar fashion in stones of jargon, or zircon, by Vauquelin's German friend Martin Klaproth in 1789, also went through a long purdah, only being isolated by Berzelius in 1824.) Later, it emerged that glucina was not the only sweet-tasting metal compound, and it was renamed beryllia and its associated element, beryllium.

For those seeking riches, news of these experiments must have come as a disappointment. Even the most precious gemstones were shown to contain no precious essence, as the more alchemically minded investigators had surely hoped they would. Unlike the dirty ores, from which gleaming metal could be extracted, these crystals lost all their value when they were processed in the laboratory. Just two years before Vauquelin's experiments with emerald and ruby, the English chemist Smithson Tennant even burnt a diamond away to nothing, proving that it was made of nothing more exotic than carbon.

The modern chemists had their reward though – Vauquelin had his chromium and beryllium, Sefström and Berzelius their vanadium, Klaproth his zirconium. Their work cleared away much confusion in the jewel trade. Tales of precious artefacts seen by excitable explorers in remote lands could now be subjected to more sceptical examination. It became obvious, for example, that many stones claimed to be emeralds were too large to be true gems, and that the term was being employed simply as an admiring simile for all manner of green objects that were actually made of jade or even glass. Today, when much progress has been made in the manufacture of artificial stones, the word 'gem' is generally reserved for natural specimens. Classification according to colour is more of a problem. Because the colour of gemstones arises from impurities in them, there is no rigorous definition of what makes an emerald or a ruby. A beryl is therefore simply a stone too pale to be called an emerald on an arbitrary scale of greenness.

Increasing colonial trade with countries rich in these minerals, such as Burma and Colombia, together with machine cutting techniques, saw to it that coloured jewels grew in popularity through the nineteenth century. Jewels were fascinatingly ambivalent in an age when strictness of morals was matched by sumptuousness of ornament. Only virtuous women and wisdom are rarer than rubies, according to the Bible. Wearing jewellery was an indication of virtue, yet also an enticement. The stones themselves are naturally beautiful, but there is devilment in the art with which they are cut, and it is no great surprise to find Mephistopheles giving Margareta a tempting casket of jewels in Goethe's *Faust*. The famous 'jewel song' of Gounod's opera version of the story amplifies this transaction as the chaste heroine laughingly imagines herself transformed into a worldly princess – in Bernstein's wicked parody of the aria in *Candide*, Cunégonde mordantly reflects that if she's not pure, at least her jewels are.

The imputation of purity is doubtless one reason why Ruby and Beryl became popular Christian names in Victorian times, only falling from favour in the 1930s. Today, Ruby may be undergoing a revival, but you have to search a little harder for other names inspired by gemstones: Esmeralda is now fashionable for girls, Jasper for boys.

The spread of precious stones as luxury consumer goods has prompted more knowledgeable allusions in literature. The emeralds Edmund Spenser refers to in *The Faerie Queene* or those in Milton's *Paradise Lost* might be any green gem, their precise hue less important than their general rarity. But we imagine that the Emerald City in L. Frank Baum's 1900 fable *The Wonderful Wizard of Oz* is truly built of that stone. And here the colour may be significant. Easily distracted academic economists have interpreted the story as an allegory of United States monetary policy at the end of the nineteenth century: the yellow brick road represents the gold standard leading the way to the Emerald City, the colour of the greenback dollar, governed by the ineffectual wizard, who is President Grover Cleveland. The allegory depends for its message on the fact that Dorothy wears silver slippers, which become the symbol of the populist 'free silver' movement that was pressing the United States mint to make silver tradeable for coin (in the same way that gold was already) following the discovery of new deposits in the American West. Having escaped notice when the book was first published and the issue was topical, this amusing undercurrent was then completely buried in the legendary film version of the story made in 1939. By this time, the subtext was technological rather than economic: Dorothy was famously given ruby slippers to celebrate the Technicolor process in which the part of the film is shot. The 'silver screen' was dead.

## The Crimson Light of Neon

Imagine finding the proverbial painting in the attic. You take it to be examined and are assured that it is an original, indeed a masterpiece, and, what's more, that it's by a painter completely unknown to the art world. Naturally, you return to the attic to see what else you can find. And in the dust, you uncover another painting, and then several more – a complete oeuvre, in fact, of a great master whom nobody knew existed.

This is what happened to William Ramsay, the professor of chemistry at University College London, who discovered five new chemical elements during the 1890s. These new elements bear a strong family resemblance: all are gases, all colourless and odourless, all remarkably unreactive. They earned the name of the inert or noble gases, and most chemists found them boring. Today, however, it is their laziness that makes them useful to us, primarily for lighting: when subjected to electrical excitation they glow brightly while remaining chemically unchanged.

Ramsay made the first of these discoveries in 1894, working

with Lord Rayleigh at the Cavendish Laboratory in Cambridge. Rayleigh had found that nitrogen obtained from minerals by chemical means was mysteriously lighter than nitrogen that remained in the air after burning all the oxygen. Ramsay solved the puzzle by burning shavings of magnesium in atmospheric nitrogen. Most of the gas combined with the reactive metal. But a little was left over, and its spectral light when excited to a glow did not correspond to any known substance. Rayleigh and Ramsay announced their discovery of a new element which they named argon; 'a most astonishingly indifferent body', they wrote. Because argon is heavier than nitrogen, its one per cent presence in air had made the atmospheric nitrogen seem a fraction heavier than the chemically made nitrogen. At a college dinner, the poet A. E. Housman proposed the toast 'argon' and called on those assembled to 'Drink to the gas.'

Ramsay became excited that argon might be the first of a group of elements that would form a new column in the periodic table. In 1895, an American geochemist wrote to Ramsay with news that he had obtained an inert gas by heating a mineral sample. Ramsay wanted to see whether this was argon, too. He scrabbled around for comparable samples, even begging specimens of a likely-looking uranium mineral from the British Museum (he was rebuffed). Soon Ramsay had repeated the American's experiment and examined the spectrum of the gas that came off. But the spectral lines did not correspond to argon. They indicated something still more unexpected, matching lines previously observed in the light of the sun. This time, Ramsay had confirmed the terrestrial existence of the gaseous element helium.

Ramsay spent the next three years trying to obtain further gaseous elements from minerals. The day when the new element might turn up became a laboratory joke, but the day never came. In May 1898, he and his assistant, Morris Travers, tried a new tack, taking advantage of new technological developments that enabled gases to be liquefied in large quantities. Since argon is relatively abundant in the air, they reasoned that other equally unreactive gases might be all around us too. They obtained a gallon of liquefied air and carefully boiled it off until just a small residue remained. Analysis of this residue once again revealed new spectral lines. They turned out to be due to a dense gas that Ramsay and Travers called krypton, a name they had first considered for argon. (Krypton means 'hidden', argon means 'lazy', so as far as the gases' chemistry goes the names are much of a muchness, but since krypton is rarer than argon it was a good call.) Ramsay telegraphed his wife, who was in Scotland, with news of the discovery. 'You get a new element every time I come away,' she wrote back, her faith in her husband's abilities clearly rather greater than his colleagues'.

Their basic hunch confirmed, Ramsay and Travers scaled up the same experiment by a factor of a thousand, starting not with liquefied air but with liquid argon. Despite the jibes of rivals and sceptics, Ramsay was confident of success. Anybody who wanted to overtake them would first have to make several bucketfuls of argon, which in itself was no small matter. In a series of careful evaporations, the men this time detected a light gas which boiled off ahead of the argon. In June, Ramsay announced this latest discovery. Ramsay's thirteen-year-old son Willie, astutely suggested the name novum for the new element, an idea that his father immediately accepted, at least in essence: 'neon' merely reflects the convention of using Greek rather than Latin roots when naming elements.

Once again, Ramsay and Travers confirmed their find using a spectrometer. Placing an electrical potential across a volume of the gas, they were delighted to see a distinctive new glow. Travers was not only an able laboratory hand; he also became Ramsay's biographer, and was not too modest to include himself as a third-person character in the narrative. His account of the day surely ranks as one of the best accounts of the dramatic moment of discovery:

As Ramsay pressed down the commutator of the induction coil he and Travers each picked up one of the direct-vision prisms, which always lay at hand on the bench, hoping to see in the spectrum of the gas in the tube some very distinctive lines, or groups of lines. But they did not need to use prisms, for the blaze of crimson light from the tube, quite unexpected, held them for some moments spell-bound.

Beginning again with liquid neon and krypton, they found one further noble gas, xenon, 'the stranger'. With evidence of these elements' uniqueness resting solely on their spectra – there were no measured physical properties and no observed chemical reactions – it is unsurprising that Ramsay had his detractors, especially as he had got into the habit of occasionally announcing his discoveries before actually making them. Not least among the doubters was Dmitrii Mendeleev, who had declared in 1895 that argon did not fit in his periodic table, and so must be a heavy form of nitrogen. The British scientists spent the next two years purifying samples of their new elements in order to prove their existence once and for all. In 1900, the sceptics were finally persuaded. Towards the end of that year, Ramsay gave a major lecture summarizing his experiments, which was then published in the *Philosophical Transactions of the Royal Society* with a quotation from Sir Thomas Browne's *Religio Medici* at its head: '*Natura nihil agit frustra*, is the only indisputed Axiome in Philosophy. There are no Grotesques in Nature; not anything framed to fill up the empty Canons, and unnecessary Spaces.' Ramsay had filled up five spaces in the periodic table, and a few years later was awarded the Nobel Prize for Chemistry, by which time others had discovered the radioactive gas radon that completed the tally of the noble gases.

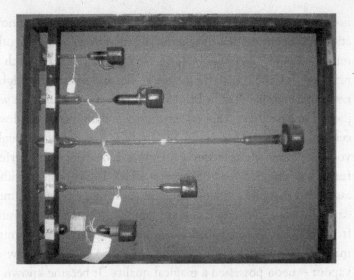

Ramsay's laboratory at University College no longer exists, but many of the gas discharge tubes he used to demonstrate their colourful radiance have been kept. Alwyn Davies, an organic chemist who has nevertheless developed an enthusiasm for Ramsay's work, leads me to an unpromising breeze-block corridor and pulls open some drawers. Inside are the dumbbell-shaped glass tubes of various lengths, blown by Ramsay himself and labelled according to the gases they contain. On the inside of the glass are smoky deposits from the vapour of the platinum electrodes, the only sign of wear. Some of the tubes, he assures me, still work.

Every element is new at the moment of its discovery, and so might deserve the name neon. Yet Travers's 'crimson light' would fulfil its destiny more completely than anybody could have predicted as the century turned.

As early as 1902, the French inventor Georges Claude began experimenting with electrical discharges within sealed tubes of neon. On 11 December 1910 he demonstrated the first commercial neon lamp to visitors at the Paris motor show. Claude's innovation was to ensure that the chemically inert neon inside the tube remained pure, uncontaminated by more reactive gases such as nitrogen that could corrode the electrodes and reduce the brightness of the discharge. The bright red light was arresting, but was judged to have limited appeal for domestic illumination, and was certainly ruled out for use in automobiles because of the high-voltage equipment needed to generate it. However, it was perfect for advertising – the activity with which it has been indelibly associated ever since. Neon light shone brightly even on sunny days and could penetrate the city smoke, making signs visible from far away. With no apparent source of the light – no burning material, no incandescent filament, merely a hovering glow of vapour – neon possessed a magical quality. It became known as 'liquid fire'.

Claude was able to make his neon tubes larger and brighter by adding substances such as carbon dioxide and using constant pumping to maintain the correct vapour pressure. Furthermore, the tubes were essentially permanent; they could be manufactured remotely, filled with gas and then transported to a building where they could simply be affixed, wired up to an electricity supply, and left to run. The world's first neon advertising message began beaming the word CINZANO to promenaders on the Champs Elysées in 1913, the same year that Stravinsky's *Rite of Spring* received its riotous premiere near by, and that musical chronicler of technological progress Erik Satie penned a little piano piece, 'Sur une Lanterne', whose optional lyrics begged the new city lights: 'N'allumez pas encore. Vous avez le temps . . .' ('Don't light yet. You have time . . .')

But modernity beckoned, and the lights went on without delay. Claude prospered, taking out foreign patents and gaining a virtual monopoly in neon tubes. Neon advertising arrived in the United States in 1923, when the media mogul and entrepreneur

Earle C. Anthony bought from the Claude Neon company of France for the reported sum of $2,400 a pair of 'Packard' signs for his Los Angeles car dealership.

Named at the outset for its own novelty, neon became the sign of the new. The cool red heat of pure neon was quickly complemented by other colours produced by different mixtures of gas. Argon-filled tubes shone a pale blue. Adding a little mercury gave a bright white light. Using tubes made of coloured glass completed the electric rainbow. 'Neon' in all its hues was curiously in harmony with the times. Paris and New York were perhaps the two cities with the greatest claim on the world's attention in the first part of the twentieth century, and both made much of the new material. The artist Fernand Léger, who was working in Paris when Claude's first sign went up there, was later excited by the ever-changing reflections of primary colours on New Yorkers' faces produced by the signs on Broadway. The rise of Art Deco, launched at the Paris Exposition of 1925, coincided with a proliferation of automobiles and the expansion of cities and their new suburbs, each of which developed their own forms of nightlife. It was not simply the shiny new style's emphasis on superficial gloss that made it a good fit with this new technology. With a rising tide of consumers in pursuit of amusement after dark, it was inevitable that neon became a characteristic feature not only of the entertainment districts of the major cities, but also of resorts from Miami to Le Touquet, whose new restaurants, bars and apartments sprouted neon signs and were sometimes even outlined in neon to emphasize the modern architecture.

What excited Léger, however, drove others to distraction. In John P. Marquand's Bostonian diary of a nobody, *The Late George Apley*, the eponymous Boston Brahmin, a man nonplussed by all things modern, is appalled by a visit to Broadway, where he sees new electric signs moving 'in nervous patterns'.

Such horrors are only to be expected of Manhattan. But when a similar illuminated advertisement appears in Boston, he launches a quixotically ineffectual campaign against it: 'nor was it due to Apley's indifference that a large electric sign, advertising a certain inexpensive variety of motor car, still flaunts itself insolently over Boston Common. He justly called this sign, to the end of his days, "Our Badge of Shame."' Marquand is making a connection here with more literal badges of shame, specifically the early New England custom of marking adulterers with a red letter A, the 'scarlet letter' of Nathaniel Hawthorne's novel. Red neon lights suggest that the city is prostituting itself to commercial interests – and worse. They advertise the globalized bazaars of Piccadilly Circus and Times Square, but also the more decadent pleasures of the Pigalle and the Reeperbahn. Though 'red-light districts' in fact antedate neon lighting by a few years, the colour association is unfortunate nevertheless for neon and may be an additional reason why its light was considered unsuitable in the domestic environment. For those who want to read it, neon provides the luminous writing on the wall of our electric Babylon.

It was not only in the worldly cities that neon found its role. With the paving and numbering of America's national highways from the 1920s, wayside gas stations, motels and diners gained vital prominence from neon. Brighter than other lights, neon signs were visible at greater distances, especially in the open expanses of the West and in the clear desert night. And if the light was visible at greater distances, then the lettering also had to be larger so that the message would be legible from afar. Roadside signs were designed to be seen a mile away – then perhaps driven past at sixty.

But in rural communities as in the cities, the newfangled glare of neon could be a signal of corruption. In *The Neon Bible* by John Kennedy Toole, the church sign depicting the Bible 'with

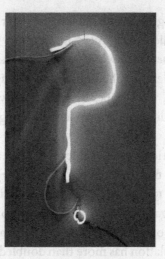

its yellow pages and red letters and big blue cross in the center'
casts a searing light that symbolizes the oppressive power of the
Mississippi preacher who hounds the boy narrator's family of
'fallen-away Christians' to death and exile.

The hovering transcendence of neon light has tempted artists
to make their own luminous glyphs. These often twist the familiar form of advertising signs to spell out more elliptical messages.
The fun is in adapting a medium that is all about instant gratifica-
tion to say something slow or mysterious. For most, the actual
craft of making the signs is an irrelevance. But Fiona Banner
makes her own glassware, a manual procedure that connects her
with the first such tubes ever made, those in Ramsay's laboratory.
'Neon best sells things for "now",' she explains to me. 'Immediate
desires – sex, kebabs, movies.' But the disembodied quality of the
light also carries timeless memories of stained glass and the sky
itself, making it both 'a retinal and cultural come-hither. When
lit, it (its physicality) is hidden in its own light, the object disap-
pears in order to be legible. It is a way of being able to say
something (a word) without any voice.' Banner's recent work
picks apart this language. *Every Word Unmade* is a set of twenty-six

separate neon signs, one for each single letter of the alphabet, the essential ingredients of urgent messages as yet uncomposed. A work called *Bones*, meanwhile, gives life to the punctuation marks that are always left out of commercial neon signs. For Banner, the glowing marks, with shapes like primitive weapons found on an archaeological dig, accrue deep new meanings.

Nowhere marries neon's wilderness entreaty and its careless urban glamour more effectively than Las Vegas. Incorporated as a city only in 1911 when it had a population of just 800, Las Vegas really began to boom in 1931, when construction began on the nearby Hoover Dam; gambling was made legal that same year. The population has more than doubled in each decade ever since and stands close to two million today. The place had a garish character from early on. The first neon beacon in the desert went up in 1929 on what was then still appropriately named the Oasis Café and was followed by the art deco tower of the Las Vegas Club in 1930, and then by a cavalcade of hotels, clubs and casinos. It was the signs ('Caesars Palace', 'Golden Nugget', 'Stardust', 'Flamingo') that defined the main commercial avenue of the city known as the Strip. With land cheap and vistas long, the signs were often taller than the sprawling buildings they advertised. But size was never going to be enough in such a competitive environment. Increasingly imaginative designs were commissioned, with flashing colours and animated graphics of, say, wine pouring into a glass or beer frothing in a stein – though seldom of cash tumbling into eager hands. The dinosaurs driven to extinction in this ceaseless parade of natural selection meanwhile are preserved in the city's 'neon boneyard'.

The restless light is all too much for Raoul Duke and his attorney in Hunter S. Thompson's *Fear and Loathing in Las Vegas*. When they check into a hotel, they find directly outside their window 'some kind of electric snake . . . coming right at us.

"'Shoot it,' said my attorney.
"'Not yet' I said. 'I want to study its habits.'"
But Duke's attorney wisely draws the curtains.

Two who did study its habits were the architectural theorists Robert Venturi and Denise Scott Brown. Following Ed Ruscha and the Pop artists, who were first to reappraise the aesthetics of the commercial strip, they decided to make Vegas 'our Florence'. (Tom Wolfe had already compared it to Versailles.) Venturi and Scott Brown noted that in many cases the light *was* the architecture. The buildings are not lit tastefully like historic landmarks; they are themselves light. They are given luminous outlines, and every surface becomes an illuminated sign for something, whether they are casinos or 'marriage chapels converted from bungalows with added neon-lined steeples'. In their enthusiastic embrace of everything vernacular in Vegas, the only thing the architects didn't like about the illuminations was their tendency to produce 'big problems with bugs'. Their revulsion may have been more than just physical: perhaps they saw the insects drawn to the light as a metaphor for our own helpless attraction to neon temptations.

But a nuisance to some is an opportunity to a serious lepidopterist such as the young Vladimir Nabokov, who once 'caught some very good moths at the neon lights of a gas station between Dallas and Fort Worth'. Nabokov's was much more than a childhood hobby, and on this same transcontinental car journey he also discovered a new species of butterfly, which he named *Neonympha dorothea* after the student who was doing the driving. Nabokov, that master of wordplay, must have loved it that the Linnaean nomenclature of his find was also able to incorporate the name of the light by which it was found.

The image of insects swarming around a neon sign was one the author employed much later in *Lolita*, his notorious novel about a Parisian émigré writer Humbert Humbert's sexual pursuit of a twelve-year-old nymphet. The later stages of the book describe a road trip around the United States, punctuated by motor courts, gas stations and candy bars. On one level the story clearly tells of old Europe's (Humbert's) infatuation with the new America (Lolita), but it's a neon-lit 1950s America that turns out to be much less innocent than it appears – for Humbert is surprised to learn as their journey together begins that his captive Lolita has already been corrupted. Finally, Humbert brings himself to free Lolita to make her own life, consoling himself with the slaughter of one of her other seducers. He drives away from the murder scene to the accompaniment of 'sherry-red letters of light' and a restaurant sign in the shape of a coffee-pot repeatedly bursting into 'emerald life'.

## Jezebel's Eyes

The Old Testament heaves with painted ladies. 'Though you enlarge your eyes with paint, In vain you will make yourself fair,' the Lord warns the daughters of Zion (Jeremiah 4:30). The sisters Oholah and Oholibah are judged for their lewdness in taking to their beds 'desirable young men' from Assyria, Egypt and Babylon. The men couldn't help themselves of course: 'they went in to her, as men go in to a woman who plays the harlot', lured by come-hither looks, jewellery, and the fact that they had gone to the trouble of washing themselves and then painting their eyes (Ezekiel 23:40).

The deeds of Jezebel, the wife of Ahab, the king of Israel in the ninth century BCE, are so unwholesome that she is dragged back to make a guest appearance in Revelation as the very embodiment of unrepentant sexual depravity. Her name has been a byword for shameless womanhood ever since. It is easy to tell she's no good for she too 'put paint to her eyes' (2 Kings 9:30). St Jerome's Vulgate Latin translation identifies the substance she used as *stibio* – antimony.

The Bible makes other mentions of antimony, as the soft setting for precious stones for example, which could be referring to any lustrous metallic alloy, but identifying the cosmetic as antimony is a safer bet (although the black powder that was in long-established usage for darkening around the eyes was in fact antimony sulphide, the element and its compounds being hard to disentangle at a time when the basic rules of chemical combination were still unknown). The Hebrew and Arabic term for this substance is *kuhl*, from which the modern eye-shadow kohl is named.

Despite the dramatic pictorial evidence of wall paintings that black eye make-up was a feature of daily life longer ago in ancient Egypt, it is not clear whether it was antimony that was being used. Certainly, there were other black powders available, the handiest being carbon in the form of lamp black or the darker bone black, which was often used to coat the eyelashes. (This 'mascara' came to be just as damnable as the black antimony eye-shadow, it seems: the word stems from the Italian for 'witch'.) But antimony was regarded as the superior product and, aside from appearing to make the eyes brighter, was claimed to produce a variety of benefits from soothing the brow to dilating the pupils, an effect due perhaps to the element's being an eye irritant.

Antimony is one of many often hazardous substances that have been enlisted over the centuries in the cause of making us more beautiful. A technical compendium called *Harry's Cosmeticology* runs an alarming gamut from aluminium (powder for glittery eyes) to zirconium (salts to strengthen the fingernails). The listing includes arsenical pyrites as a depilatory, bismuth oxychloride as a pearlescent addition to lipsticks, and cadmium sulphide for fighting dandruff, in an index that counts more than forty of the elements in all.

I rush to my wife's dressing table to see what lurks among the sweetly perfumed, innocuous-looking white creams, but am surprised – and alarmed – to find that, unlike foods, the packages carry no explanatory labels. Has a business with an infamous record of using dangerous chemicals so cleaned up its act that no accounting is necessary? Or is the risk judged acceptable in the name of beauty? Although chemists have devised new materials that offer marvellous colours, the cosmetics industry finds it prudent to confine itself to a relatively small repertoire of dyes approved by bodies such as the United States Food and Drug Administration. So-called interference pigments then tweak the few basic colours in order to produce the greater range of shades that the market demands. Today, many lipsticks use intensely coloured organic dyes such as fluorescein dispersed in a medium of white titanium dioxide powder rather than heavy metal pigments. Odd plasticky additions supply other desirable effects, such as microscopic balls of Perspex used to give a pearlescent gloss.

Samuel Johnson possessed 'an apparatus for chymical experiments' and called chemistry his 'daily amusement'. His familiarity with the science is reflected in his famous dictionary, which includes entries for most of the elements known to science by the mid eighteenth century, including the newly isolated cobalt. His entry for antimony is especially entertaining. 'The reason of its modern denomination,' he suggests, as opposed to the Latin *stibium*,

is referred to Basil Valentine, a German monk; who, as the tradition relates, having thrown some of it to the hogs, observed, that, after it had purged them heartily, they immediately fattened; and therefore, he imagined, his fellow monks would be the better for a like dose. The experiment, however, succeeded so ill, that they all died of it; and the medicine was henceforth called *antimoine; antimonk*.

It was natural that Johnson should think to include antimony. What today seems to us a rather marginal element was then held in high esteem, having been regarded as centrally important by the alchemists. Though the dark arts of alchemy had begun to yield to a more systematic chemistry, the alchemists' texts were still necessary references, even if they were not always quite what they purported to be. A mysterious tome called *The Triumphal Chariot of Antimony* speaks of antimony's ability to cure leprosy and the French pox, but also contains some solid science, accurately noting the element's two contrasting forms – a brittle silvery metal and a grey powder. Alchemists regarded this duality as significant, for it brought antimony close to both mercury and sulphur, mother and father of all the metals.

The fact that antimony can take these two forms was a source of much hermeneutical head-scratching. To complicate matters, the element occurs in nature most often as the sulphide stibnite. Prepared in certain ways this black powder, Jezebel's kohl, changes again, turning orange, with no hot furnace or special apparatus required to shift between these confusing forms. Johnson was having a little fun with his etymology: the word antimony is derived in fact from *anti monos*, meaning against singleness, a straightforward reference to these shifting properties, and nothing to do with the element's ill effects on the brothers of the church (though the word monk comes from *monos* too).

The supposed author of *The Triumphal Chariot*, Basil Valentine, and his fellow alchemists regarded the amorphous grey phase of antimony – 'the sage's matter' and the 'grey wolf of the philosophers' – as the tantalizing last stage before the realization of the philosophers' stone because of its ambivalent capacity to produce either the lustre or the hue of gold, but never yet the two together.

More alluring still is the metallic form of antimony, which has been celebrated since antiquity for its ability to solidify in a large crystalline mass that combines the gleam of precious metal with the faceted symmetries of gemstones. The phenomenon was undoubtedly noticed when the pure element, or regulus, was first made. The disc of crust that formed on top of the melt became known as 'star antimony' after the characteristic radiating pattern produced as the antimony crystallized in the cooling vessel.

Isaac Newton, an alchemist as much as a mathematician and physicist, read Valentine and followed his recipe for making the regulus of antimony, thinking he might be able to use its gleaming surface in telescopes. One biographer invites us to believe that the stellate pattern he produced may have helped him to visualize the lines of force that led him to develop the theory of gravity. This seems fanciful to me. I can see how the patterns might inspire ideas about optics, which Newton was also investigating at the time of his experiments with antimony, but not gravity. I decide to go in search of antimony stars.

For such beautiful artefacts of nature, they are surprisingly hard to find. I am quickly disabused of the idea that every Victorian collector would have owned one and that they would today be littering the store-rooms of provincial museums. However, photographs and illustrations show crystal patterns that are not acicular, which is to say like the chrome-spoked wheels of a sports car, converging on one central point, as would have to be the case for a diagram of gravitational force. Instead, the solid rink of antimony tends to be divided up into polygonal domains, smoother and larger near the centre of the disc and disintegrating into extravagantly foliate intaglio, like frost on a window, towards the outer edges according to its rate of cooling. The overall design is indeed star-like, not in the astronomical sense of a point source from which all light radiates, but in the manner of a child's typical drawing of a star with a number of triangular points or, more particularly perhaps, like Renaissance emblems of the flaming sun.

Perhaps this resemblance inspired another famous experiment with antimony. In 1650, one Nicolas le Febre was demonstrator in chemistry at the Jardin du Roi in Paris, where, for the edification of the young King Louis XIV, then a thoughtful eleven-year-old boy, he embarked upon 'the Solar Calcination of Antimony' by means of 'Magical and Celestial Fire, drawn from the Rayes of the Sun by the help of a refracting or burning Glass'. Le Febre focused sunlight on to 'the Stellat or starry Regulus' and showed that the product of the reaction weighed more than the antimony he had started with. Perhaps the antimony star gave the young Louis the idea for the symbol that would shine over his long reign as the Sun King. Whether it did or not, the experiment was a milestone in modern chemistry for demonstrating proper method in place of alchemical obscurantism, and showing the first glimmerings of the understanding that the air itself contains chemical elements.

# Part Five: Earth

## *Swedish Rock*

At the beginning of my chemical odyssey, I had traced a map of the world and placed a dot on it wherever one of the elements had been discovered. A very curious map it turned out to be. Aside from zinc and platinum, which were found without the assistance of Western science in India and the Americas respectively, all the dots relating to the naturally occurring elements fell in Europe. A cluster of dots in Berkeley, California, accounted for most of the elements heavier than uranium that have been made artificially following the discovery of nuclear fission. Another cluster at Dubna, north of Moscow, showed where some more recent radioactive elements have been synthesized.

Europe showed four major hotspots of earlier date – London, boosted by the multiple successes of Davy and Ramsay, and Paris could claim just over a dozen elements apiece. Berlin, Geneva and Edinburgh also made their mark. But the two largest clusters of dots after Paris and London were both in Sweden, one at the old university city of Uppsala, and the other in the capital, Stockholm. Swedish science could claim the discovery of at least nineteen elements, more than a fifth of the naturally occurring total. Many of them, indeed, celebrate the places where they were found (yttrium, erbium, terbium and ytterbium named after the ore mine at Ytterby; holmium named after Stockholm itself) or after more or less romantic ideas of Scandinavia (scandium, thulium).

(In old Europe, it is quite often the case that the elements are named for places associated with their discovery. For instance, strontium happens to be the only element named after a place in the British Isles, Strontian in Scotland. It's more often the other way round in the United States, where chemical knowledge helpfully preceded westward expansion and the rush to uncover the riches of the wilderness. America's Golden Hills and Silver Lakes were no idle poetic allusion; they expressed a direct connection to the earth into which adventurers hammered their tent pegs, and the hope, whether fulfilled or ultimately dashed, that these precious metals were to be found there. Aside from gold and silver, a dozen elements appear nakedly in town names, from the expected Irons, in Missouri and Utah, Leadville in Colorado, and Copper Center, Alaska, to the frankly surprising Sulphur, Oklahoma, Cobalt, Idaho, Antimony, Utah and Boron, California.)

What made Sweden figure so largely in the story of the elements? I have been concerned throughout this book to suggest that we are familiar with many of the elements in a cultural way, that is to say without ever stepping into a laboratory. We know neon and sodium for the light they cast, iodine for its brown balm, chromium for its cheap shine. Others, such as sulphur, arsenic and plutonium, we tend to know more by reputation. The Swedish finds are mostly obscure by these measures – they include metals such as manganese and molybdenum, and a good handful of the elements known collectively as the rare earths, a group that includes all of those elements directly named after Swedish locations. They have not made a mark in the sense that they have become a byword for some human horror or delight. Yet these elements too have a cultural connection, and, as their toponymy implies, it is a connection that runs deep. Paris and London revealed new elements to the world because they were major centres of intellectual life. Berkeley and Dubna happened to be the sites chosen for the specialist machines needed to manu-facture the heavier elements beyond uranium in the periodic table. But in Sweden's case, the logic is unshiftable: her elements sprang from the very ground of the country.

In order to learn more about this fertile womb of the elem-ents, and how it was that there were men of science ready there to act as midwives, I decided to visit Sweden myself in order to find out how it was that two cities on the margins of Europe – one scarcely more than a small town – sustained an advantage over a period of a century and a half that enabled them to out-strip London and Paris in this race of discovery. During the first half of the seventeenth century, Sweden briefly became the new superpower of northern Europe, overrunning Nor-way, Finland and parts of Russia, northern Germany and the area now occupied by the Baltic states. Underwriting the expansion were Sweden's vast reserves of iron and copper ore, which provided the necessary military and economic might. In time, these imperial ambitions were superseded by a new and more attractive idea, that of Scandinavia. But the mining con-tinued, and it was from these mines, during the years of Sweden's gentle decline, that this one nation made its lavish contribution to the periodic table. I muse on this history and how it is reflected in the Swedish elements, whose names became gradually less localized with each successive discovery, from yttrium in 1794 to scandium in 1879, as my plane flies over lakes and forests towards Stockholm.

In Stockholm, I meet Hjalmar Fors, a young historian of chem-istry with a wispy blond beard, who has agreed to lead me on a walking tour of the scientific sights. We start in the Stortorget. It means big 'square', although it is a little square on the little island of Stadsholmen, Stockholm's old town. Commanding one short side of the square is a red-washed merchant's house with baroque gables and a psalm inscribed on a tablet above the door; it is where Carl Scheele, the nearly man of oxygen and chlorine, worked as a pharmacist around 1768. Our next stop is the national mint, which stands on the waterside hard by the Royal Palace. Here, in 1735, Georg Brandt, the Guardian of the Mint, speculated that the blue colour of the smalt ore obtained as a by-product from the royal copper mines might be the clue to a new element. The Board of Mines based at the mint was responsible for the analysis of minerals and maintained the first chemical laboratory in Sweden long before there was one at the university in Uppsala or anywhere else. This laboratory had been around for long enough to have fallen into some decay when Brandt came on the scene and set about modernizing it. It seems that Brandt may not have been thanked for this labour, however. Brandt was a rationalist, but his masters were Rosicrucians disinclined to let go of their mystical beliefs. In time, however, Brandt gained greater control over operations and was able to exert a more enlightened influence. During the later part of his career he continued to devote much energy to disproving charlatans' claims to have transmuted silver and other metals into gold. It took him seven years to obtain the first specimen of cobalt metal. It was, Hjalmar tells me, the first truly modern discovery of a chemical element, that is to say the first to be supported by a solid notion of chemical theory, not merely alchemical hocus pocus.

We move on, crossing over the water to Karl XII Square. Among the grand buildings overlooking the greenery is a substantial nineteenth-century yellow-ochre edifice that was the headquarters of the iron mines when Sweden was the world's greatest exporter of the metal. An extended bas-relief frieze runs around the top of the building. It shows heroic figures engaged in every stage of the iron-making process, from extracting the ore, to smelting in the furnace, to casting the pigs. Lower down, the façade is punctuated by plaster medallions of Scheele, Berzelius and other great Swedish chemists. 'I rather like this,' says Hjalmar with a glint in his eye. 'Nobody really knows who these guys are now. But there they still are up on the wall.'

I am beginning to sense the vital connection not only between national prosperity and mining, but also between mining and chemical science. Sweden's first true chemists were kept in employment one way or another by mining interests. Unlike their peers in Britain and France, they tended to be highly trained in the analysis of minerals. They worked at the Royal Mint or the Board of Mines or in direct collaboration with mine owners. They obtained their specimens from the mines and were frequent visitors themselves to the mines at Falun and Västmanland, a day or two's ride from Stockholm or Uppsala. Here they were doubtless to be seen scrabbling through the tailings for exceptional stones or searching the exposed veins for

glimmers of unusual colour, often performing their preliminary analyses in makeshift laboratories set up on site. These were no aristocrats amusing themselves in plush home laboratories, but realists aware that wealth came by virtue of hard labour out of the cold earth, and that any scientific knowledge gained in addition to that would be the more meritorious if it led to an increase in that wealth. And these men were justly rewarded for their realism by the business community whom they served: no medallions of Lavoisier or Cavendish are to be found on the commodity exchanges of Paris and London.

We stop for a beer at a café in the Royal Hop Garden public park, with Scheele's unreliable statue gazing down on us. Hjalmar tells me of his pipe-dream to rewrite the history of science by shifting the centre of gravity eastward to focus on the intellectual traffic across the Baltic, between the Scandinavian, German and Russian powers, ignoring for a change the bickering rivalry of the English and the French. It is a project that could restore to their rightful place in the chemical pantheon the shrinking violets of Sweden's mining laboratories, men whose congenital modesty led them fatally to delay publication of their discoveries or even to shun it altogether, so helping to ensure they never received their due on the world stage – Johan Gahn, the discoverer of manganese; Torbern Bergman, the *éminence grise* behind many of the metals first isolated from Swedish sources, but the direct discoverer of none; and Scheele, who, having moved on from Stockholm, found even Uppsala too exciting, and spent the years when he might have been famous holed up in the little Västmanland town of Köping, batting away offers of employment from wealthy English and German patrons.

The next day I take the train to Uppsala. Stockholm was the commercial and financial centre where the metals from up-country were assayed, traded and made into coin. What was Uppsala's role? Uppsala has the oldest university in Scandinavia, established in 1477, but it wears its history lightly. The place hardly seems like an intellectual hothouse. There is life on its few shopping streets but no real bustle. Pedestrians and bikes weave happily around each other, and cars are few: it is easy to picture the city as it must have been two or three centuries ago. A fast-flowing river in a granite channel divides town from gown, but students are as thin on the ground as the shoppers.

I meet Anders Lundgren, a lecturer in the history of science at the university, who sports a tangle of grey beard on a scale that Hjalmar Fors can only aspire to. As we stroll along, I observe idly what an extraordinarily genial place Uppsala seems to be. 'Yes it is,' Anders agrees. 'Now. But not in the winter.' It is early June. He points out a white dormered building where in the mid eighteenth century Uppsala's first professors of chemistry, Johan Wallerius and Torbern Bergman, made their laboratories. It was in this building and its successors that most of Sweden's element discoverers either learnt their art or passed it on as professors to the next generation. This was the case whether they came from Stockholm, like Anders Ekeberg (tantalum) and Per Cleve (holmium and thulium), or from the mining country, like Brandt (cobalt), or even from the Finnish territories like Johan Gadolin (yttrium): they all spent time in Uppsala. Peter Hjelm (molybdenum) and Lars Nilson (scandium) were two more Uppsala graduates. Scheele, meanwhile, ran the pharmacy in the town square where he made the first chlorine and oxygen, although he played no official part in academic life. Uppsala has a splendid university museum, the onion-domed Gustavianum, but, in what I am coming to accept as the inevitable Swedish fashion, it omits to celebrate even one of these men.

Equidistant from Stockholm and the mines, Uppsala was the third point of a triangle – the thinking brain to the labouring hand and the pumping heart of the Swedish body politic. It was not a straightforward relationship, however. The crown needed the mines to fund its imperial ambitions, and the mine owners no doubt enjoyed the royal patronage. But it is not immediately clear why either of them needed the scientists. Anders Lundgren has studied the way mining influenced the development of science in Sweden. 'Chemistry could never pay back to mining,' he explains. The miners had no need of chemists to point out the valuable ores to them, and may well have resented these secular intruders with their blithe unconcern for the miners' obscure traditions. And if the chemists were lucky enough to discover new elements, then these were of no interest either. They might fill odd gaps in theoretical understanding. 'But theories of chemical affinity were no use in a blast furnace.'

Yet the chemists did gain support such that for a long time in Sweden chemistry was probably the only science to offer the prospect of a decent career. The crown gained intellectual prestige from supporting the laboratory at the Board of Mines, and mine owners emulated this largesse on their own modest scale. Some mine owners indeed, such as Berzelius's patron and collaborator Wilhelm Hisinger, were scholars in their own right. The national 'minerography' that Hisinger produced at the age of twenty-four, for example – a kind of atlas of mineral resources – was less the stake-claiming project of an avaricious prospector, far more a product of humanistic pleasure in knowledge for its own sake.

Though it may do little to commemorate them, the Gustavianum did contain one further clue to the extraordinary success of Sweden's chemists. I have indicated elsewhere that the discovery of elements often relies on the possession of some special technology, and that given this device or that technique, the discoveries then often come in a rush. It is easy to believe that in the eighteenth century there was no such technology in place to assist in prising the rare earths and other elements from the grudging Swedish rock. This is so in the sense that no high-powered gadget suddenly unlocked the floodgates, and the discoveries came instead in painful fits and starts over a long period. Yet during all this time there was one tool which no self-respecting Swedish chemist was ever without: his blow-pipe. The specimen in the museum is perhaps twenty centimetres in length and appears to be made of iron. It is essentially a thin, elegantly tapered tube not unlike a cigarette-holder. At one end it is slightly flared so as to produce a good seal with the user's mouth. At the other, the airway is bent through ninety degrees and forced through a small hole, while a separate outlet drains off spittle as in a musical wind instrument.

This simplest of equipment was the key to the analysis of unfamiliar minerals. It had the great advantage that it could be used in the field. According to a guide published by one distinguished Swedish mineralogist, it amounted to nothing less than

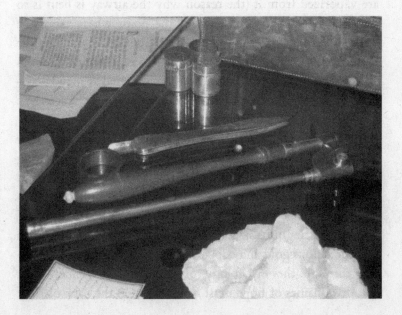

a 'pocket laboratory'. Even Goethe, a keen scientific amateur, had Berzelius instruct him in its use. The blow-pipe was eventually superseded by the spectroscope, but it remained a feature of analytical chemistry teaching until the middle of the twentieth century; Anders Lundgren remembers trying one out at school, and describes to me how it works. Though simple enough, it demands powerful lungs and fiendish skill if it is to deliver good results. Its great versatility lies in the fact that it may be used to blow a jet of air through different regions of a flame, thereby producing a high-temperature zone that is able either to oxidize or to reduce (the reverse chemical process) a mineral sample positioned in the way.

If all the senses are alert, a vast range of diagnostic information may follow from this apparently crude process. If the user has the breath to sustain the flow of air for ten or fifteen minutes to allow the sampled mineral to come to red heat, the colour of the flame may change repeatedly as different metallic elements are vaporized from it (the reason why the airway is bent is so that the user gets a clear view of the point where the flame strikes the mineral). The smell of the vapours can confirm the presence of non-metal ingredients such as sulphur, selenium and tellurium. Even the sound the mineral makes may be significant, with a crackling noise, for example, being characteristic of chemically bound water being freed from the sample.

The blow-pipe seems to me to express the essence of what Anders describes to me as Sweden's characteristically 'boring good chemistry'. Even the scientists may have been bored at times, sweating and puffing in frustration over unreadable minerals, dissolving them up to produce an endless succession of almost indistinguishable salts. It is a world that seems far from the miraculous gold and copper, amber and jewels, that scintillate through the mythology of this land. I wonder what colourful flames of hope must have danced at the back of their minds as these men performed their inexorable experiments. This was science of the green-fingered kind that relies on crafts-manlike skill, immense patience and intimate familiarity with its raw materials. It was these qualities, more than mercurial brilliance or extravagant equipment, that explained the discovery of so many of the elements in this north-eastern extremity of the European continent. These qualities, and of course the prodigal abundance of the soil.

## Europium Union

The rare earths are not rare, but they are unsung. This group of elements to which so many of the Swedish discoveries belong populates a row of the periodic table which is usually shown dangling below the rest of the table like a 'vacancies' notice slung below a motel sign. Its members are: scandium, yttrium, lanthanum, cerium, praseodymium, neodymium, promethium, samarium, europium, gadolinium, terbium, dysprosium, holmium, erbium, thulium, ytterbium and lutetium. And, rare though they are not, you can be easily forgiven for never having heard of a single one of them.

Nor are they really gritty 'earths': all are middleweight metals. It is only because they for so long resisted extraction from their oxide ores that they earnt this general label. Recalcitrance may be the rare earths' main unifying characteristic. In other respects, their properties are finely differentiated; indeed it is a matter of chemical semantics as to whether some of them – scandium and yttrium at the beginning of the sequence, and lutetium at the end – even belong on the list at all.

In almost every case, the isolation of the rare earths – from yttrium in 1794 to promethium in 1945 – was a punishing grind. These discoveries do have the distinction, however, that they were made (apart from the anomalous radioactive promethium) by thoroughgoing chemists. They were not dependent upon some unique technology closer to physics as was the case with some other groups of elements – the alkali metals discovered electrolytically by Davy, Ramsay's inert gases glowing in their discharge tubes, the transuranium elements thrown together in the Berkeley particle accelerator. The separation of the rare earths was chemistry all the way. The typical procedure was to dissolve an ore in acid to form a solution containing a mixture of salts. This was then slowly evaporated so that the salt of each element crystallized out in turn, while the overlying liquid retained others in solution. Careful repetition of this process – sometimes thousands of times – enabled chemists eventually to separate these very similar substances one from another, and from them then to isolate the new elements at their heart. It was, as one chemical historian drily remarks, 'an enormous undertaking which would have difficulty attracting grant support today'.

Monotonous as it undoubtedly was, this long-running project was heaven to a particular type of experimental mind. The Swede Carl Mosander boasted of his ignorance of chemical theory, and showed exactly how unimportant it was by discovering more rare earths than anybody else by dint of sheer hours spent at the laboratory bench. There is a definite rustle of the anorak about these elements. With the hindsight of established scientific knowledge, it might be easier to crystallize their stories in words than it was to crystallize the elements themselves, but it would quite probably be just as tedious as the original exercise. So I shall not spend time on all of them, but rather single out one or two as representative of the set. The differences between them are slight in any case. They behave in generally similar ways and do similar things. Some of these things are useful – rare earths are widely if sparingly employed in ceramic glazes, fluorescent lamps, television screens, lasers, alloys and refractory materials – but the matter of choosing which one of them to use in many of these applications is, if not entirely inconsequential, then at least somewhat arbitrary. Not always, however. Occasionally, one of these rare earth elements recommends itself for the job above all the others.

If you take a €5 note and hold it under an ultraviolet light, the dull yellow stars that slice through the Classical arch on the front face of the note suddenly glow an intense red. On the reverse, a three-tiered Roman bridge appears to float in a ghostly greenish light above a river of indigo. This light comes from special inks incorporated into the notes in order to make them difficult to counterfeit, which are excited to luminescence by the powerful ultraviolet irradiation.

The exact nature of the chemical compounds used is of course kept secret by the European banks. However, in 2002, just months after the euro went into circulation, a pair of Dutch chemists decided to amuse themselves by performing an unusual spectroscopic analysis. Freek Suijver and Andries Meijerink at the University of Utrecht shone ultraviolet light on to euro notes and recorded the exact colour tones of visible light they emitted as a result. From this they were able to declare that the red light was due to ions of the rare earth element europium bound in a complex with two acetone-like molecules. They were less sure about the other colours, but speculated that the green might be due to even more elaborate ions involving europium combined with strontium, gallium and sulphur, and the blue to a europium complex with barium and aluminium oxides. They suspended their enquiries at this point, warning others who might be tempted to follow them that 'any further investigation into what causes the luminescence of euro notes would constitute a violation of the law'.

But the unravelling of this little secret hardly gets to the nub of the matter. What we would really like to know is how it came to be decided that, of all the many inks that perform this trick, it should be inks based on europium that were chosen. It was, after all, a political decision in the end that a bank note issued in the name of European unity should have its mission slyly reinforced by impregnation with a chemical element named in celebration of the very same idea.

Europium metal is as soft as lead and must be stored under oil to stop it bursting into flame in air. It is the most reactive of the rare earths, and because of its urge to bind strongly to other elements it was among the last of them to be discovered.

In Art Nouveau Paris, Eugène-Anatole Demarçay began to suspect that samples he had acquired of samarium and gadolinium – the future europium's next-door neighbours in the periodic table, discovered a decade or so before – might not be pure. Demarçay was a gaunt, severe-looking man whose chief glory was his florid moustache. He spent his early career working in the laboratory of a noted Paris perfumer, but soon went freelance and acquired renown as a spectroscopist – he could read the spectrum of a substance like 'the score of an opera', according to a contemporary. (The Curies would soon come to him to confirm their discovery of the elements polonium and radium.) Beginning in 1896, Demarçay prepared salts from his samarium and gadolinium samples and, through the exhaustive process of separation by crystallization, was able to isolate a new salt that was progressively richer in an unidentified substance. By 1901, he had amassed sufficient evidence to confirm his suspicion that this was a new element.

Demarçay named his element for the entire continent of Europe, but he seems to have left no account of why he did so. His choice ran conspicuously against the contemporary trend of naming new elements after nation states. Not long before, Mosander in Stockholm and various others at the University of Uppsala had seen to it that a clutch of new elements were given names after places in Sweden. Gallium was named for France in 1875; germanium for Germany in 1886. Freshest of all in the memory for Demarçay was the Curies' discovery of polonium in 1898, in which he had assisted. Perhaps all this nationalistic fervour was reason enough to cast an opposing vote.

In the Europe of 1901, prescient souls had long since begun to suspect that nation states might not be everlasting, with Frenchmen taking the boldest line. Victor Hugo was the first to speak of a 'United States of Europe', in 1848. The Breton philosopher Ernest Renan dared to ask in a famous lecture given in 1882 at the Sorbonne 'What is a nation?' and to imagine that 'They will be replaced, in all probability, by a European confederation.' This cosmopolitan spirit was evident at the Paris Universal Exposition of 1900, where more than fifty million people came to see forty nations drawn from all the continents exhibiting their wares – including specimens of the newly discovered rare earths.

Most European citizens, it has to be said, showed no sign of such ideals, and nationalism, having successfully delivered the unified new states of Italy and Germany, set off on a downward spiral based less on liberal highmindedness and more on ethnic and linguistic tribalism. Before long, it seemed that any group of self-asserting Ruritanians, to use the historian Eric Hobsbawm's term, might suddenly decide to call themselves a nation. For Demarçay, a well-travelled autodidact accustomed to forming his own ideas, it must have been easy to resist the prevailing drift of nationalism and nail his colours to the mast through his chemical discovery. He would have surely welcomed the advent of the European Union and rejoiced to see his metal become part of its economic fabric.

The European Central Bank seems incapable of spreading this joy, however. It wilfully misunderstands my request to know who fought for europium and drearily asks for my 'understanding that, for security reasons, we cannot comment on the chemical components of the euro banknote security features'. I know what the chemical components are; what I want to know now is who was the wag in the Brussels bureaucracy who made sure it was europium that was used. The bank requires its money to incorporate these security features, including raised print, metallic strips, watermarks and holograms, but does not actually print the stuff itself, and therefore does not specify that europium or any other particular material must be used for the luminescent dyes. So others may have been responsible in any case. However, the leading printers of euro bank notes will tell me nothing either.

I reread Suijver and Meijerink's paper and see that it contains a clue. Seeking confirmation of their europium revelation, they contacted the Dutch National Bank, and were eventually put in touch with a researcher there. During the course of their conversation, the bank employee accidentally let slip something that jogged a memory in the Utrecht chemists. 'A few years earlier he and a colleague visited our laboratory,' Meijerink remembers. 'During his visit we were able to supply him with a great deal of information about luminescent materials. Not surprisingly, he could not give us much information.' So were the Utrecht chemists actually responsible for planting the idea of using europium in the first place? Did they simply stage their analytical 'discovery' in order to lay a false scent, or did they do it because they couldn't resist claiming their own paternity, as it were, of the europium dyes in the euro? Or, alternatively, was it the mysterious visiting bankers who had the brainwave, bowing to fate when they heard that an element called europium was one of those that would be suitable for the job? For now, nobody seems to want to claim this inspired decision as their own.

## Auerlicht

The girl is naked from the waist up; and from there down she is draped only in the lightest gauze. She is kneeling, with her head cocked to one side, and smiles naughtily out from beneath her brunette curls. In her right hand she appears to hold a dazzling halo of

white light, at the centre of which an even brighter light shines out – 'appears' because the light has no obvious source or connection; it is pure illumination. She grips the stem of a large sunflower for support and is framed by vigorous tendrils of other growth. Set to one side in front of the plane of the picture – it would be anachronistic within it – is a standard gas street lamp. The message becomes clear. This vestal virgin is holding forth the promise of a new light, like the light of the sun, that will brighten the world.

Giovanni Mataloni's 1895 poster advertised the improved gas lighting of the Brevetto Auer company of Rome ('guardarsi dalli contraffazioni' – beware of imitations). It was one of hundreds of similar images that appeared in cities across Europe and America around the turn of the century. Full-colour illustrated posters were the latest fashion in advertising, and no field of commerce was more assiduous in seeking the public's favour than the fast-expanding domestic lighting industry, where gas and electricity vied ceaselessly with rival innovations.

The breakthrough that enabled gas to maintain its advantage over the newfangled electric lighting for a little longer during the closing years of the nineteenth century was made by Carl Auer, later the Baron von Welsbach, a Viennese who had completed his studies in Heidelberg with Robert Bunsen, long the guru of European chemists. On his arrival in Heidelberg in 1880, Auer showed the great man a modest collection of rare earth mineral specimens that he had amassed, and Bunsen set him to work analysing them, laughing off Auer's protests that the quantities were insufficient. This project set the course of his career; and the rare earths would make his fortune. Auer's annus mirabilis came in 1885 back in Vienna, when he succeeded in separating the supposed element didymium into two true elements, which were duly named praseodymium and neodymium. Their green and pink compounds make them attractive for use in ceramic wares and in tinted glass for protective eyewear.

Auer was not content merely to add to the number of the rare earths. In his Heidelberg days, he had marvelled at Bunsen's already famous burner, with its tunable flame that could be adjusted to simmer or roast. He had noticed how, when turned up high, the bunsen burner flame would cause his rare earth ores to glow brightly with their own light. He began to explore this phenomenon with different combinations of metal oxides. It was well known that a flame set against a piece of lime (calcium oxide) will produce the incandescence known as limelight. Auer's investigation included the oxides of magnesium and beryllium, both closely related to lime, as well as those of his rare earths and other elements.

Gas lighting was well established in streets and homes by the mid nineteenth century, but the light it shed was limited by the luminosity of the flame it produced, which depended in turn on the mixture of hydrocarbons being burnt. Candles and oil lamps gave a brighter light than gas, but only gas could be supplied continuously. Auer believed that a lamp design in which his rare earth oxides were positioned close to the gas flame might yield a brighter light. Over a period of several years, he soaked sleeves of cotton mesh in different mixtures of rare earth and other salts. Once dried, these sleeves or mantles, now stiff with encrusted oxide, were placed around the flame, which burnt away the fabric to leave a brittle lace of the refractory oxide. This would then glow brightly in the heat of the flame.

Little was known about the properties of many of the oxides, and still less about how they behaved in combination, so there was no way of predicting which composition would produce a white incandescence. Auer first patented a gas light with a mantle made of a mixture of magnesium, lanthanum and yttrium oxides in 1885, but its fragility and sickly green light meant that it remained unpopular. However, by 1891, he found that thorium and cerium oxides mixed in the proportions of ninety-nine to one gave a satisfactory white light (thorium is not a rare earth but is the heavier – and, unknown then, radioactive – cousin of cerium). Mantles made of this material were more sturdy and quickly caught on. Unusually for a scientist, Auer was an astute businessman, and his name soon became more widely known even than Bunsen's. For while the bunsen burner had its place in the laboratory, the bright new Auerlicht, as it was known, was of utility to all, and was rapidly distributed across a grateful continent by various Auer companies. Some 90,000 Auer mantles were sold in Vienna and Budapest alone in 1892; twenty years later, annual production stood at 300 million units.

It can have done no harm to the inventor's prospects that Auer, a variant of the prefix *Ur-*, is an archaic German word for the dawn. At the exact moment when the first of Auer's bright gas lamps was lit outside the Opern Café in Vienna on 4 November 1891, Auer's countryman Gustav Mahler, no stranger to the place, was composing a song that would be incorporated into his second symphony, called 'Urlicht' – primordial light.

Auer clearly acquired a taste for appending his name to his inventions. He followed up the success of the gas mantle with an osmium electric filament, the Auer-Oslight – even as he perfected his gas mantle, Auer was hedging his technological bets by experimenting with materials for the electric lamps that he already suspected might one day replace them. In 1903, he patented an alloy of cerium and iron – he called it Auermetall No. 1 – that produced sparks when struck. 'Flints' of this material are still used in cigarette lighters today. Everything Auer touched seemed to turn to light. No wonder that on the occasion of his ennoblement he chose for his coat of arms the motto 'Plus Lucis' – more light.

Cerium is the most abundant of the rare earths, and is more plentiful than many familiar elements such as copper. It seems destined to remain widespread if largely unappreciated in our lives. The metal is used to improve the performance of cast irons, steels and aluminium alloys. Known as jeweller's rouge, its powdered oxide is a fine abrasive used to polish gemstones and glass. In the nineteenth century, it was recognized that cerium salts were anti-emetic, and they were also incorporated into cough tinctures, anti-bacterial treatments against burns and tuberculosis – conveniently for medicines these salts also possess a characteristic sweet taste. More recently, there has been excitement at the discovery that cerium oxide added to diesel fuel greatly improves its combustion efficiency. And it is still used in illumination to brighten the powerful lights used on film sets.

Cerium was the discovery of the greatest Swedish chemist of them all, Jöns Jacob Berzelius. Unlike some of his more bashful compatriots, he did publish his results in a timely fashion, as well as maintaining lively correspondence with his international peers and receiving chemical pilgrims at his laboratory. If he has been written out of the popular history of science, then the blame for it lies squarely with prejudices to the West.

The mineral world was not Berzelius's first love. Born in 1779, he came of age at a time when it was already thought the glory days of Swedish science were over. The talented apothecary Scheele was dead along with the mineral chemists Brandt and Gahn, who had identified new iron-like metals in the ores of the royal mines. So too was the world-renowned botanist Carl Linnaeus, who had dared to think that man might classify all of nature and had made a good start on the job with his binomial nomenclature for plants and animals.

Trained as a physician, and intrigued, like many scientists of the time, by the effect of electric currents on living organisms, Berzelius wanted to know the secret of life. In order to learn this, he would first have to discredit fashionable theories of vitalism and offer a more rational explanation for animal and human physiology. One helpful step forward was to call the field 'animal chemistry'. For a brief period at the beginning of the nineteenth century, this became a hot topic in science. An 'Animal Chemistry Club' formed as a special interest group of the Royal Society in London counted Davy among its regulars, with Berzelius being an active corresponding member. But the scientific problems proved largely intractable. The challenges posed by the chemistry of life nonetheless honed Berzelius's skills as an analytical chemist, and he attracted the support of the prosperous mine owner Wilhelm Hisinger. Despite his avowed distaste for inorganic chemistry, Berzelius had little choice but to respond, like so many Swedish scientists before him, to the call of the earth.

Berzelius was responsible for the introduction of now famil-iar items of laboratory equipment such as rubber tubing and filter paper but, unlike Bunsen with his burner or Davy with his miner's safety lamp, he failed to pin his name to them. He intro-duced concepts and words which have since proved too useful to restrict to the scientific lexicon: 'catalysis' and 'protein' are his neologisms. He did invaluable work on the proportions in which elements and their compounds combine with one another, which underpinned the theory of atoms advanced by the English Quaker John Dalton, and for the first time gave chemistry a solid quantitative foundation. It was Berzelius too who saw the need for an abbreviated notation for the elements and invented modern chemical symbols. His system of a one- or two-letter code, often based on the element's name in Latin, has since become iconic far beyond the discipline of chemistry. Putting together these last two ideas – the symbol for each elem-ent, and the understanding that they combine with one another in fixed proportions – led inevitably to the first chemical formu-lae, those concatenations of letters and numbers that mean everything to chemists and simply appear random to the rest of us. ('Ah, $H_2SO_4$, professor!' is Flanders and Swann's impression of how scientists greet one another in their satirical treatment of C. P. Snow's polemic concerning the 'two cultures' of the arts and sciences.)

This system of notation appears to us now both familiar and alienating. Its advent in 1811, however, was a graphic revelation. The consequences for the scientific comprehension of matter were far-reaching. In their modern laboratories, the alchemical quest now left firmly behind, Enlightenment scientists had begun to show that they could synthesize simple compounds found in nature – Lavoisier had combined the gases hydrogen and oxygen to produce only water; the exotic flammable metals that Davy had isolated could be burnt to recreate the oxides found in naturally occurring minerals. Berzelius's system finally erased any lingering distinction between the essence of a material obtained from natural sources and the same material produced in the laboratory. Once a substance such as ammonia, say, is identified as $NH_3$ rather than 'spirit of hartshorn', it is suddenly clear that it no longer matters where it has come from in order for it to be what it is.

This would be enough to secure any chemical reputation, and yet there is more. For Berzelius was also the discoverer not only of cerium but of three more chemical elements – thorium, selenium and silicon, all elements by their nature tightly earthbound. All of these discoveries relied on his intimate involvement with mining and industry. The silicate minerals from which he eventually extracted pure silicon provide Sweden's bedrock. He found selenium, an element related to sulphur, in the sediment of a sulphuric acid plant in which he had an investment. Thorium and cerium he isolated from unusual mineral specimens sent to him for examination. In the case of cerium, in particular, Berzelius worked closely with his patron Hisinger in Stockholm, as well as at Hisinger's country estate, and at the mines themselves, systematically electrolysing various salts derived from the specimens, which had been obtained at one of Hisinger's abandoned mines. Berzelius chose the name cerium, inspired by the recent discovery of the dwarf planet Ceres, and following the precedent established with uranium and Uranus a few years before.

Although the Swedes were the first to use electrolysis in the effort to obtain new elements, they struggled to gain rightful recognition of their priority in this over Davy. When the French chemist Vauquelin learnt of this work, he commented that, if the Institut de France had known of it in time, Berzelius would have shared the Napoleon medal that it awarded to Davy.

Berzelius may have suffered from the airbrushing of chemical history due to the later achievements of the Germans, French and British, but I felt that Swedish reserve was not helping his case. I had come to Stockholm partly in the hope that I

might get a glimpse of the chemicals Berzelius had collected and labelled with his compelling new notation. I had seen these in a colour plate in an old biography – small vials with chunky glass stoppers or corks filled with dusts in pastel shades of blue, yellow, grey and soapy green, each with its identifying formula in Berzelius's own handwriting. One container of candy pink stood out puzzlingly from the rest – few salts are really pink. The caption implied that these treasures were on display at the Berzelius Museum. But the museum no longer stands, and its contents, I am told, are held in storage crates at the Royal Swedish Academy of Sciences awaiting the day when his successors see fit once again to honour his mighty contribution to the inventory, theory and language of chemistry.

## Gadolin and Samarsky, Everymen of the Elements

In 1788, Carl Axel Arrhenius, a Swedish army lieutenant and mineralogist (the latter interest acquired while learning how to test gunpowder in the laboratory of the Royal Mint), discovered a black, asphalt-like ore aggregated in the flesh-pink feldspar of the Ytterby mine. Arrhenius was excited by the thought that it might be a source of the dense metal tungsten, discovered a few years before. He promptly sent a specimen of it for analysis to his friend Johan Gadolin, the professor of chemistry at the university in Åbo (now Turku in Finland, then part of the Swedish empire). After a long delay, Gadolin responded with more interesting news: the lieutenant had discovered the ore of a new rare earth element. Gadolin named the ore yttria after the Ytterby mine, and worried what this latest find might mean for chemistry as a whole. 'It is not without great trepidation I dare speak of a new earth because they are right now becoming far too numerous,' he wrote, 'for it seems to me rather fatal if each of the new earths should only be found at one site or in one mineral'.

Gadolin's fears that the rare earths would proliferate turned out to be well founded. This one Ytterby mineral would in the end reveal not one rare earth element, but four, and their apparently exclusive association with the location of their discovery made it seem right to name each of them after it: yttrium, erbium, terbium and ytterbium. Later, Per Cleve separated the oxides of two more new metals from the same ore, and named them, more broadly, holmium after Stockholm, and thulium after the old name for Scandinavia, Thule. Meanwhile, in a different Ytterby mineral, Anders Ekeberg discovered another new element – a metal, but this time not a rare earth – tantalum. By 1879, the Ytterby mine was finally the source of seven chemical elements in a list that then totalled seventy in all.

The mineral from which Gadolin obtained his yttria, called ytterbite to begin with, was soon renamed gadolinite in his honour. However, this was not to be his only or his greatest claim to scientific immortality. For later, the element gadolinium was, with samarium, the first to be named, not after a figure in mythology, nor after some Greek neologism based on its chemical behaviour, nor even after the place where it was found, but after a real person. Samarium was discovered in 1879, and named for a Russian mining engineer, Vasili Samarsky. Gadolinium was identified the following year.

It was not until 1944 that a new element was again named after a person. That was curium. Other new arrivals followed this honorific convention during the 1950s, including einsteinium, fermium, mendelevium and nobelium. All these elements esteem scientific figures already much esteemed for their achievements. You might think these elements are all rather remote from daily experience. At any rate, they seem much less familiar than the figures for whom they are named. With gadolinium and samarium, it's surely the other way around: their discoverers are even more obscure than they are. Although you may not have heard of these two metals, they are both more abundant than tin and are found in every modern home. Gadolinium is used in magnetic recording discs and tape, while the miniaturized loudspeakers of personal stereos depend on highly magnetic alloys of samarium. Who, then, were Gadolin and Samarsky, this pair who sound like a Milwaukee firm of attorneys? And who was it that wished to praise them by rendering tribute in this uniquely enduring form?

Johan Gadolin was born in Åbo in 1760 into a family that included two bishops of the city. Deviating slightly from the custom among clerical families of gentrifying one's name into a Latin form (like Linnaeus), Johan's grandfather had taken the name Gadolin, meaning great, from the Hebrew. Gadolinium would thus become the only element with its etymological root in Hebrew. His investigation of the black mineral sent by Arrhenius was the nearest Gadolin came to discovering an element. In 1827, his collection of minerals was lost when fire destroyed Åbo and its university; yttrium metal was finally isolated by others the following year. Gadolin lived into his ninety-third year, long enough to savour the honour of having the mineral gadolinite named after him though not long enough to see the arrival of gadolinium.

Vasili Evgrafovich Samarsky-Bykhovets rose to the rank of colonel in the Russian Corps of Mining Engineers. Stationed in the southern Ural mountains in 1847, he noticed an unfamiliar friable mineral the colour of burnt caramel, which he was curious enough to have sent to Berlin for expert assessment, where a German mineralogist confirmed its novelty and recommended the name samarskite, following the convention in the field; samarium duly followed. Little more seems to be known about Samarsky, who made no further contribution to science.

Compared with the Curies, or with pioneers such as Berzelius or Lavoisier or Davy, seemingly destined never to feature in the periodic table, Gadolin's and Samarsky's contributions seem minimal. Why were these two so favoured? If their achievements are not recommendation enough, we must turn for our answer to the later investigators of the elements that came to be called samarium and gadolinium.

In 1879, Paul-Emile Lecoq de Boisbaudran, the wealthy son of a Cognac distillery owner, extracted certain salts of a rare earth element thought to be didymium from a sample of Urals samarskite. When he combined the salt solution with another reagent, he found that it did not produce the single precipitate he expected, but formed a sediment with two distinct phases. 'Didymium' was not an element at all, but a complicated mixture of unknown rare earths. Separating the two residues, he was able to show that one of them was a compound of a new element, which he named samarium. The following year Jean Charles Galissard de Marignac in Geneva, working with a different specimen of the 'didymium' mineral, isolated another new rare earth oxide. Lecoq confirmed de Marignac's discovery and suggested the name gadolinium for this new element. (Then, five years later, Carl Auer finished 'didymium' off for good by showing that it contained two further true elements, neodymium and praseodymium.)

So it was Lecoq who was responsible for shooting these relative nonentities to stardom in the periodic table. What was his motive? As we have seen, the last quarter of the nineteenth century was the zenith of European nationalism. Should he not have named samarium instead after France or Paris, where he worked, and gadolinium after Geneva or Switzerland, where his friend Marignac was based? In fact, he was probably wise not to try, for he had already shot his bolt in this direction, and done so in spectacularly controversial fashion.

Lecoq had made his first contribution to the periodic table in 1875 when he isolated a new element from zinc ore. He presented a specimen of it to the French Academy of Sciences and named it gallium in honour of France. The trouble began a couple of years later when suspicions were raised that the naming was not quite the patriotic gesture it seemed, but was in fact Lecoq's sly way of naming his discovery after himself – though the Latin for France might be Gallia, the Latin for *coq* was also *gallus*. The controversy was such that Lecoq was obliged to deny that he had chosen the name in self-homage. The episode would have been painfully fresh in his mind as he worked on the didymium minerals.

After the embarrassment of gallium, it is possible that Lecoq simply wished to play it safe. And nothing was safer than following the accepted naming of the source minerals as closely as he could, replacing the -ite suffix of the geologist with the -ium of the chemist. It seems that he chose samarium because it was obtained from samarskite and gadolinium because it was obtained from gadolinite with no more ado. If so, it is chemistry's loss. Many more minerals are named after geologists than elements after chemists, and not only because the list of minerals is long compared with the list of chemical elements. Mineralogists have a long and fine tradition of naming minerals after pioneers in their field, a practice which self-deprecating chemists have by and large been loath to emulate. As a consequence, many chemists who never got their name attached to an element nevertheless have a mineral named in their honour. Among these, cleveite, tennantite and wollastonite honour chemists who discovered elements. Gadolinium and samarium are two rare examples of the favour being returned. Gadolinium must stand as the memorial for all the chemists who have struggled to free a new element from its mineral source, and samarium for all the mineralogists who spotted that unusual mineral in the first place, chipped it from the native rock and brought it to the attention of the world. Neither Gadolin nor Samarsky are the greatest representatives that might have been chosen for this duty: they are the everymen of the elements.

## Ytterby Gruva

Hearing the stories of the rare earths, I felt I was beginning to understand more deeply where the elements came from. Of course, I knew that in their totality they came from the earth, the sea and the sky. I wanted to penetrate beyond this obvious syllogism – everything is made of elements, so the elements are found everywhere – and identify a kind of *locus classicus* for these fundamental ingredients of all matter. After all, they are universal only in a sense. True, everything is *made* of elements, yet the pure elements themselves seem oddly elusive, almost always locked away in inscrutable minerals and compounds. Searching for the elements in nature was like raiding a bakery and finding plenty of cakes and buns but no sign of the flour and sugar from which they were made. You do not find nuggets of aluminium or rivers of mercury when you go for a walk in the country. Still, I thought, there must be places where the aura of the elements could be felt.

It was time to visit a mine. I did not want to go to the Great Copper Mountain at Falun, the vast mining centre celebrated by E. T. A. Hoffmann, founded in the thirteenth century and still in commercial operation as recently as 1992. Nor did I want to go to Hisinger's mines in nearby Västmanland. Berzelius and Hisinger had discovered cerium from ores dug there, but they had been searching for Gadolin's yttria, the ore that took its name from the village of Ytterby, whose little mine gave the world not only yttrium but six other elements besides. I wanted to go to this most prolific source of the elements.

Ytterby is the site of what is said to be the oldest feldspar and quartz mine in Sweden. It lies on the island of Resarö, one of an infinity of rocky islands east of Stockholm where Sweden disintegrates into the Baltic Sea. In the early eighteenth century, the feldspar quarried here went for making porcelain in Swedish Pomerania, while the unusually pure quartz was sent to Britain for making glass. But to the element collector, it was only when men examined the impurities that impeded these operations that the mine revealed its real treasure.

If Ytterby is a place of pilgrimage, who are its pilgrims? The mine closed in 1933. But chemists and mineralogists have continued to seek it out. In 1940, Brian Mason of the Smithsonian Institution in Washington DC found the mine partly flooded, although there were still large blocks of pegmatite, the feldspar quartz that bears the black, triangular-faced crystals of gadolinite, lying around. A few years later, he went back and was disappointed to find the site had been taken over and fenced off for use as an oil depot and there was no longer any public access. In his account of his visits, he lists twenty-five minerals, which between them contained quantities of yttrium, tantalum, niobium, beryllium, manganese, molybdenum and zirconium as well as more common mineral elements such as aluminium and potassium.

Mike Morelle, the schoolmaster responsible for inculcating my own fascination with the elements, came across the mine by chance in 1960, when he was invited to stay at a business colleague's holiday home in Ytterby. Walking in the woods near by, he found himself in a rocky pit that reminded him of the V2 rocket crater that had appeared outside his bedroom window one morning in 1945. The slopes were overgrown, and there was no obvious mine entrance, but he noticed a few signs of former quarrying activity. Only later did he learn from his host of the chemical significance of the site.

Jim Marshall at the University of North Texas and his wife, Jenny, have toured the world seeking out the sites associated with the discovery of each element in a vacation project that has turned into a ten-year obsession. Their objective is to visit every relevant mine, laboratory and chemist's home. The idea took root when the Marshalls reached the end of the line in a genealogical investigation that had given them a taste for purposed European travel. What better way to continue in the same vein than to contrive a project that would oblige them to visit some of Europe's most pleasant cities and a smattering of rugged out-of-the-way locations, with each site's inclusion always justified by its connection with the overall mission? The itinerary would be long enough to be daunting, but sufficiently contained to be achievable in the end. Naturally, they have visited the great cities such as Paris, Berlin, London, Edinburgh and Copenhagen, and also obscure spots such as Strontian and the gloomy Transylvanian mine where tellurium was first found. The Marshalls' 'walking tour of the elements' seems to offer the perfect touristic marriage of the urbane and the sublime, so that, on the gallium trail, for example, they find themselves both in Lecoq de Boisbaudran's Cognac and in the misty Pyrenean mountains where he obtained the zinc blende from which he extracted the element. Sadly, though, as their written and photographic records of these trips attest, they often find these sites unmarked, neglected or built over. Jim and Jenny finally reached Ytterby in 2007.

On my own journey, I have been asking the chemists and the historians of science I meet whether they too have been to Ytterby. Few have. Andrea Sella was taken on a ferry cruise across the Gulf of Bothnia to Sweden while attending a conference in Gadolin's city of Turku, but he bunked off the proffered side trip to Ytterby in favour of a day's sightseeing in Stockholm. Even my local guides, Hjalmar Fors and Anders Lundgren, have not made the excursion to the birthplace of so many elements.

In art and literature, the studio and the writer's desk retain a certain fascination. But it does not matter where Newton and Einstein were when they revolutionized the laws of physics; it merely matters that they did so. You can visit the family home in Lincolnshire to which Newton retreated when the plague swept through Cambridge where he made his most important discoveries. In the garden is an apple tree which, it is tentatively claimed, is grafted from the one that famously dropped its fruit on to the great man's head. But it offers no insight into Newton's revelation concerning the law of gravity; it is just an apple tree. I hoped Ytterby would be different. Here, after all, it was not the chance presence of some human genius that made it significant. This was not Stratford-upon-Avon or Dove Cottage. The meaning had to lie *in* the place, in the unique material constitution of one patch of geography.

The sky is a pale grey and the trees are dripping from recent drizzle as my bus snakes its way through the Stockholm suburbs along sliproads cut into pink and grey rock. Soon everything manmade seems to arise from this geology – the aggregate that surfaces the road, the steel barriers along its edge, the metal siding of the industrial estates, the rough-hewn stone and ochre stucco of the more imposing buildings, the red clapboard of the houses (called *Falu röd* after the Falun mines, whose cuprous ores are used to make the pigment). Everywhere, rounded boulders thrust their way through the late spring vegetation, as if it is they that are alive and burgeoning and will soon overwhelm the grass and bushes, and not the other way around.

As the bus surges on, I think about how chemistry seems to have become an almost clandestine activity. The alchemists are discredited and cold in their graves, yet the science of the elements seems to have gained little respect or respectability. Chemistry's heroes and heroines are neglected. The subject is increasingly taught hypothetically in schools, with experiments often no longer performed either by pupil or by teacher, but merely described or seen on a DVD. Chemicals are things to be feared, the necessary ones kept in their place under the kitchen sink (and designated 'chemicals', as if the sink itself and its contents are not chemicals too). I'd struggled to get hold of the simple substances and apparatus I needed for my own modest experiments; I'd visited a fireworks factory hidden behind a hedge in a layby with no company sign to advertise it; I'd heard from academics driven out of their urban laboratories to remote waste land in order to do their experiments. It seemed an odd way to increase scientific knowledge and spread understanding. The elements – many of them – are obtainable if you know where to look, but this knowledge itself is made to seem dangerous, as if it is only to be gained at the price of knowing some secret code: sulphur is to be found at the garden shop; magnesium at the ship's chandler; antimony at the artists' supplier. Surely the universal elements should belong to all humanity.

The bus crosses a couple of inlets and drops me off. I am the only one to disembark. The drizzle starts again, and I now understand why the walking map I have bought in preparation for the final stage of my journey comes in a plastic sleeve. I had hoped for a journey of epic dimensions, and am a little dismayed to find that Resarö is these days comfortably within commuting distance of Stockholm. The map shows the settlement of Ytterby and an angular G for 'gruva' – mine – at the tip of the island. I trudge along for a mile or so against the rain. The field-fares are making excited ratcheting noises in the trees. Wild geraniums grow in the verge. Soon, rock-strewn meadows give way to an idyllic suburbia, and the rain eases. The sound of children playing fills the air. Houses and gardens appear with little vegetable patches dotted with blueberry bushes and onion flowers. Blue-and-yellow pennants skip merrily in the breeze from tall poles in many of the gardens.

I follow signs to a café which I find in a boatyard. The café is no more than an open-sided hut with a handkerchief of decking looking out over the water. The paper napkins are weighed down with a piece of the pink stone. I ask the owner if he knows of the Ytterby mine and its hoard of elements. He does, but hasn't been there himself. 'I've only lived on Resarö for five years. I'm not the explorer type.'

I walk on past a street called Yttriumvägen and know I must be getting close. A little further on, two boulders of the pink rock have been placed at the side of the road. A scree path leads steeply up between them through birches and pines. To one side

of the path stand the metal posts of a sign that has been removed, but a small plastic badge stapled to an adjacent tree announces that I have arrived at a 'Natur minne'. The path is pure white and rose quartz as if in a fairy tale. I scramble up. At the top, I find a lumpy vertical expanse of rock as big as the front of a house. Ytterby gruva. The rock is grey and pink and white and black. At the foot of this little cliff the remains of some-body's campfire have been safely ringed by stones of the various colours.

The site scarcely resembles a mine of any sort. There are no workings, no spoil heaps, not even any detectable opening into the earth. It is too compact to be called desolate. The landscape is picturesque rather than ravaged or scarred. Wild strawberries cling to the crevices in the rock. I wonder if this can really be the womb that gave birth to so many elements – for yes, in the lore of the miners who once laboured here and everywhere, the earth is indeed the mother, and the ores embryos growing in her belly that they must help to bring into the world.

Looking more closely, I begin to notice signs of human inter-ference – a line of holes drilled into the rock by miners who never came back to prise away the next block, and here and there iron spikes and eyes driven into the cliff side that once sup-ported the gantries used to transport the stone down the hillside. Yet there seem to be just these few metres of worked rock wall. The rounded boulders lying near by are completely undis-turbed, with the same glaciated polish they acquired thousands of years ago. Climbing up on one, I gaze out over the treetops at a vista of small islands receding endlessly into the sea. Now something like exaltation steals upon me. I feel the roundness of the world and all its substance beneath my feet.

Time to start hammering. I have brought with me a magnify-ing glass and a small but powerful magnet to test any specimens I find. But I am ill-equipped to attack the exposed rock. My pathetic tapping on the quartz cliff – 'hard, rebellious quartz', Mark Twain had called it – gives me an immediate sense of the huge labour involved in this branch of mining. It is far harder than coal, and entirely unrelieved by softer material. As well as hammers and chisels – and later Alfred Nobel's dynamite – the miners used fire and ice to crack the stone, piling up great timber bonfires against the rock face and then dashing it with freezing water. I scan the ground for fragments broken off by winter frosts to add to my haul. I pick up a clean piece of white quartz and a piece each of the pink, grey and black rock. Then my eye is caught by a glittering trail of what I take at first to be snail slime (the slugs are out in force on this damp day), but turns out to be tiny shards of another mineral. I locate the fissure from which the pieces are falling, and find it stuffed with fragile thin plates like filo pastry which break off at the same angle like art blades. The surface of each layer has a bright, tinny shine. I have never before seen anything so obviously metalliferous coming straight out of the ground.

After a couple of hours rummaging about, I have amassed what I consider to be a representative selection of minerals, including quartz, feldspar, a grey sulphurous-smelling rock and a promising blackish stone noticeably denser than the rest, which is iridescent like anthracite, but is clearly metal-bearing.

On the shore near the mine there is a little jetty from where the minerals were shipped around the Baltic Sea and beyond, and where, doubtless, curious specimen hunters landed in the days before commuter buses. The local rotary club has erected a sign commemorating the observant lieutenant Arrhenius who started the scientific gold rush that led to the naming of six elements after this place. The area is now hemmed in by smart holiday homes, none more than 100 years old, most rather newer. I try to imagine it in Arrhenius's day, with the

noise of quarrying ringing through the pines, competing with
the squeal of the gulls. Despite the rocks and the undergrowth,
it cannot have been wild even then. It was an easy landing by
boat, and the mine was a quick scramble up the hill. I paid my
visit on a cool day in June. What was the place like in February
when the easterlies blew in across the sea from Russia? Perhaps
it was bitter. Or perhaps there was comfort in the camaraderie
of the mine and some shelter from the wind, even a little
warmth to be gleaned when the fires were lit against the rocks.
What makes this place special is not sublime landscape or the
adventure of getting here. It is something immediate and corporeal. It is the substance of the soil, the rock revealed in its
naked variety, and the knowledge that so many of the elements are native right here – uniquely as it was once thought and
as Gadolin feared, although now only emblematically so. This
earth, for me, is the source of all the elements and of our
understanding of them, the *fons et origo* of all the varieties of
matter.

I leave Resarö with my mineral haul and walk on to the
adjoining island of Vaxholm, a genteel resort dominated by the
sixteenth-century fort that squats on an islet across a narrow
channel from the town. The fort contains a small display about
the Ytterby mine, with historic photographs and, I am pleased
to see, specimens of yttrium, erbium, terbium and ytterbium
supplied by Max Whitby's elements clearinghouse. Clustered
proudly round the little glass bottles of the metals, like parents
round their children, are their source minerals – pyrrhotite,
biotite, anderbergite, allanite, chalcopyrite, molybdenite so soft
it can be used as a pencil, and uranium-rich fergusonite, the scars
made by its radioactive emanations visible on the surrounding
feldspar like tiny sunrays scratched into the mineral surface. I
worry that none of these specimens looks much like the little
stones I have scavenged, and also that I might now be carrying

dangerously radioactive material. The display includes a glossy black lump of the famous gadolinite, which certainly doesn't look like anything I found at the mine. The photographs were taken in 1893 during the mine's heyday, and show quite a major operation, with tunnels, timber buildings and rails laid for ore wagons. I had seen little surviving evidence of this industry. I learn that the site, once owned by the celebrated Rörstrand porcelain company, is now protected by the state as a 'geological treasure'. Too late, I read that it is forbidden to collect the minerals.

I am eager to learn more about my trophies when I return to London. Somewhere, I fancy, must be a mineralogist who can take one glance at my handful of stones and tell me all about them, like the wine taster who is able to name not merely the region and the year of a wine, but the vineyard and even the slope where the grapes were grown.

I take them first to Zoe Laughlin, the friend who opened my ears to the sonic characteristics of the elements with her tuning-fork experiment, who runs a materials library at King's College London. In the few days since my return from Sweden, some of the specimens appear to have changed. The sulphurous aroma has dissipated from the grey stone, and the flaky mineral with the tinny shine now looks more transparent, like sheets of cellophane. Zoe tells me it is mica (the Swedish word for it is *glimmer*). She passes a Geiger counter over the specimens in turn; to my relief, they raise barely a crackle. Nor do they reveal any fluorescent ingredients when viewed under ultraviolet light. This rules out uranium in the ore, which would glow brightly under this illumination. The preliminary surmise is that I am unlikely to have bagged any gadolinite or other minerals rich in rare earths.

Now I need a mineralogist's opinion. The Natural History

Museum runs a service – miraculous in these days when everything must turn a profit – which permits any member of the public to walk in and request an analysis of unusual minerals they have found. Minerals curator Peter Tandy hoicks his spectacles on to his head and begins to examine my stones. Most of his customers, he tells me, are people who believe they have found a meteorite (they are almost always wrong). One time, he found himself genuinely puzzled by a silvery lump of metal that somebody had brought in with just this hope, until a colleague took one quick look at it and recognized it as the remains of a wartime Italian hand grenade and called the bomb squad. He identifies most of my specimens at a glance, and takes them away for analysis by X-ray diffraction, which will identify which mineral species they are from the crystal structure. A few weeks later, Peter has disappointing news for me. I've not brought back anything noteworthy at all.

I am sorry, of course, that I have not come away from Ytterby weighed down with yttrium and the half-dozen other elements first identified in its rocks. But then, as I say, I am not a natural collector. My aim in this book has been to show that the elements are all around us, both in the material sense that they are in the objects we treasure and under our kitchen sinks, but also around us more powerfully in a figurative sense, in our art and literature and language, in our history and geography, and that the character of these parallel lives arises ultimately from each element's universal and unvarying properties. It is through this cultural life rather than through experimental encounter in a laboratory that we really come to know the elements individually, and it is a cause for sadness that most chemistry teaching does so little to acknowledge this rich existence.

We should cherish and celebrate our necessary involvement

with the elements. We may not wish to start our own periodic table, but we should at least try to be happier about the unavoidable fact that we depend in one way or another upon almost all of them. The scientist and environmental activist James Lovelock once said he would be willing to store all the high-level waste from a nuclear power station in a concrete bunker on his land. But perhaps we should spread it around: we should *all* have a little piece of spent uranium to keep in the garden as a memento of our reliance upon it for our energy.

Too much? Maybe. But what of all the other elements? The copper that invisibly brings the electricity generated by the nuclear reaction of that uranium into our homes? The rare earths in the phosphor screens of the devices brought to life by this electricity? What of the carbon and calcium that engrave all human history with their black and white? And what of the other elements that colour our world? First and last, our dependence on the elements is biological, as we are reminded when we review the sodium salt content of a TV dinner or pop a supplement pill containing selenium – the latest, by the way, in a long line of elements to be singled out as a fashionable nutrient. We eat them or avoid them, dig them up or bury them, but we rarely stop to appreciate the elements for what they are.

In the last, unfinished book that he had hoped would be his masterpiece, Gustave Flaubert invents two bumbling auto-didacts, Bouvard and Pécuchet, who decide to try their hand at every intellectual speciality the modern world has to offer. They investigate chemistry first in their unsatisfying sampling of the modern sciences, and are dismayed when they realize that they themselves are constituted of the same universal elements as all matter: 'they felt a kind of humiliation at the thought that their persons contained phosphorus like matches, albumen like egg whites, and hydrogen gas like street lamps'.

They have characteristically got it completely wrong. It is the matches that contain *our* phosphorus and the street lamps *our* hydrogen, not the other way round, and we should feel a kind of thrill.

# Epilogue

In 1959, Tom Lehrer provisionally concluded his catalogue aria of the elements (then 102 in number): 'These are the only ones of which the news has come to Ha'vard / And there may be many others, but they haven't been discarvard.' Ten more have joined the list since then. It is unlikely that they will gain the cultural traction of their forebears. They are superheavy, radioactive and short-lived, and will never find ordinary applications. They must be made in such tiny quantities that there can be no question of characteristic colours and smells. But, like all the elements before them, they are universal, they are ours; they belong to us as much as the oxygen we breathe. They belong in the periodic table too, at least insofar as they add to the sequence denoted by atomic numbers. And yet, synthesized rather than discovered, made rather than found, they seem to count rather differently.

I wondered what it felt like to bring one of these strange entities into the world. Often, I've now come to realize, something revealed at the moment of its discovery will set the course of an element's career in our culture. Chlorine's bleaching power was appreciated at the outset; so were the rainbow colours of cadmium. But these new elements, so frangible and fleeting in their existence, could never hope to worm their way into lives like this. In many ways, they must remain unreal, even to those who first make them. Could the exhilaration felt by these discoverers be anything like that felt by William Ramsay and Morris Travers when they stood 'for some moments spell-bound' and watched neon first give forth its 'blaze of crimson light' or by Davy as he danced round the laboratory in ecstasy at the fiery sputtering of

potassium? Did today's scientists believe their elements were
equal to these colourful performers? Unfortunately, they have
failed to leave the compelling accounts of their predecessors: in
order to get an answer to my question I would have to ask it
directly.

I wanted to know, too, how far the periodic table *could* go. If
it were on my bedroom wall now, should I make allowance with
rows of empty spaces to accommodate future elements? If so,
how many? One or two? A dozen? Hundreds? Not long before
he died in 1999, Glenn Seaborg, who first found plutonium and
the string of radioactive elements that follow it, gave a lecture in
which he showed a table running up to an unnamed element
number 168 – half as far again as science has managed to push
the inventory in 300 years. Was this just fancy, an old man's
hopeless dream? Seaborg seemed to disown the prospect even as
he spoke: 'We'll do well to just go on up another half-dozen elements or so,' he said. But then why show the image? Perhaps it
was his way of reminding the audience that scientific discovery
has a habit of changing the rules as it goes along. When the first
'true' chemical elements were identified, there was no way they
could be construed as numbers five and six to complement Aristotle's long-established foursome, earth, air, fire and water.
They overturned that system completely and demanded a new
one. When he drew up his list of thirty-three elements in 1789,
Lavoisier likewise had no means of knowing how many others
still lay undiscovered in their ores. In the mid nineteenth century, when the rate of discovery dropped away for a while
almost to zero, some chemists may have begun to assume they
now knew all the elements there were to know – but then the
adventure was set in motion again by the invention of the spectroscope, which allowed many more to be identified by their
characteristic flames. Dmitrii Mendeleev, despite making allowance for new arrivals, was still shocked to learn about the

existence of the inert gases and the first radioactive elements. They slipped into his periodic table easily enough, even though he fought against their inclusion at first. Will Mendeleev's design continue to be so accommodating? Or will some new element one day be found that exhibits such outlandish behaviour that the whole table must be broken up and remade?

Who can tell me what it feels like to discover an element today, and how many more such discoveries there are likely to be? For this, I need to track down the surviving discoverers of the newer elements and their successors who are still trying to find yet more, for building the periodic table is a continuing project. Although I trained as a chemist, I am mildly shocked to find I do not know their names. Cosmologists and geneticists familiar from over-exposure in the media, yes. But these chemical pioneers, no. One reason is that they are so thin on the ground. This is not simply due to the attenuated rate of discovery – down from an element every year or two through most of the nineteenth century to one every three years in the twentieth – but also to the fact that elements now tend to be discovered in batches by a few groups of researchers. This leaves fewer winners to claim the glory even if they are so inclined.

Seaborg's colleague and successor, and the only individual who can rival his record, is Albert Ghiorso. He joined Seaborg's team at its wartime base in Illinois as part of the Manhattan Project in 1944 and by 1971 could claim to be the co-discoverer of the elements numbered 95 to 105, including lawrencium, rutherfordium and dubnium. When it came to element number 106, Ghiorso was moved to ask his mentor what he thought of the name 'seaborgium'. Seaborg – who can hardly have thought the day would never come – pronounced himself 'incredibly touched. This honor would be much greater than any prize or award because it was forever; it would last as long as there are periodic tables. There are just over a hundred known elements in the universe, and only a handful of these are named after people.' Ghiorso still has a desk at the Lawrence Berkeley National Laboratory at the age of ninety-three. I write to him with my questions. But he does not respond.

After seaborgium, the laurels for the discovery of the following six elements go to the Institute for Heavy Ion Research at Darmstadt in Germany. The senior scientist at the centre during the 1980s and 1990s when these discoveries were being made was Peter Armbruster. This time, I am in luck. When I speak to Armbruster, however, he shrugs off any glory. '*I* didn't discover them. I always worked with a group.' But he surprises me by revealing that discovery can still be pinned down to that primal moment when something new irrupts into the senses. In 1981, he and his team of nuclear physicists were trying to make element number 107. The laboratory was still using noisy printers rather than silent computer screens to display results. As the equipment recorded the disintegration of the short-lived atom, 'we heard a burst of clicks'. Were those clicks any less marvellous than new light in a spectroscope?

The synthesis of these superheavy elements is in principle a matter of simple addition. Uranium is the heaviest naturally occurring element in the periodic table. Seaborg and Ghiorso had made the next elements in the sequence by bombarding targets of uranium – and then plutonium, americium and so on – with much lighter particles in the hope that some of them would stick and so form a still heavier new element. The difficulty with this – and it grew all the time – was that the targets themselves were already unstable. This increased the chance that the bombardment would simply produce a shrapnel of small high-energy fragments and no heavy atoms at all. Armbruster's breakthrough was to see that if he used atoms of certain middleweight elements as his missiles, he could go back to using stable targets. Element number 107, bohrium, was made by firing chromium atoms at a target of bismuth; 112 by forcing together atoms of lead and zinc. Because the new element survives for a few seconds at most before decaying, it must be detected not by direct observation, but by measuring the energy of its decay particles and determining the composition of the stable nucleus left behind. From this information, it is possible to calculate the atomic number of the new element that must have existed during the brief moment before the decay. Discovery in these cases is not a matter of eureka moments and apples dropping on heads. The pleasure is more like that felt by the archaeologist who is able to work out from a few shards what an ancient pot must have looked like.

Though the explorers of this far region of the periodic table are physicists, they share the chemists' urge to describe their new elements and to make compounds from them. They are motivated in this not by some soft-headed nostalgic wish to follow in the footsteps of earlier element discoverers, but by sound scientific principles. Armbruster's group has succeeded in making compounds such as bohrium sulphate and hassium tetroxide, working with just a few atoms of these elements. This, however, has been enough to prove their chemical analogy with the elements directly above them in the table, thereby demonstrating the continued validity in these uncharted waters of Mendeleev's organization of the elements. 'There were speculations that Mendeleev's table might break down with heavier elements,' Armbruster explains. 'The effect of relativity on inner-shell electrons moving at close to the velocity of light would mean that ordinary quantum mechanics would no longer apply. But we found that hassium really behaves like iron and that element 112 is like mercury.'

I ask Armbruster about names. The naming of elements is a chemists' preoccupation, the nuclear physicist points out. Adding a proton to an atomic nucleus transmutes it into a different chemical element one unit heavier; adding a neutron merely converts it into a heavier isotope of the same element. To the physicist, it seems unfair that it is only the former that warrants a new name. Notwithstanding this, Armbruster has been involved in many naming decisions. Until 1992, he tells me, it was the discoverer's right to choose a name, but this has changed in response to the priority disputes of the Cold War so that claimants are now only permitted to put forward names as suggestions. I detect a slightly shamefaced tone when Armbruster excuses his team's naming of elements 108 hassium (after the federal state of Hesse) and 110 darmstadtium. The official justification is that it completed a nesting geographic set along with europium and germanium (Darmstadt, Hesse, Germany, Europe), and was a fitting response to Seaborg and Ghiorso's earlier naming of americium, californium and berkelium (an event which prompted the *New Yorker* to quip that if only the researchers could now find universitium and ofium their work would be complete). 'This bad tradition was established by Berkeley. We wanted to do it for Europe,' says Armbruster. It seems that nationalism begets nationalism. But there is a more subtle patriotic subtext here too – an historical reassertion of Germanic strength in nuclear physics. For Armbruster's favourite of the six elements he has helped to name is number 109, meitnerium, named after the partly Jewish Austrian physicist Lise Meitner. Working in Berlin, and then in Stockholm and Copenhagen in exile from the Nazi regime after 1938, Meitner was one of the discoverers of nuclear fission, the process whereby atomic nuclei split to release massive amounts of energy. (Her work also demonstrated why the elements heavier than uranium could not be stable.) Meitner achieved this in the face of Nazi persecution and discrimination against women wherever she went. 'I was convinced that she was a very important part of nuclear physics in the twentieth century,' says Armbruster. 'And she had all the disadvantages you can have.'

By chance, I speak to Armbruster a few days after he has submitted his proposal for the name of element number 112 to the International Union of Pure and Applied Chemistry, the organization in charge of ratifying chemical nomenclature. IUPAC requires each new element to have a name that is easily pronounced and a memorable chemical symbol. The choice for 112 has apparently been narrowed down from thirty suggestions, some Germanic, some Russian, reflecting the make-up of the team that did the research. Their previous discovery, element number 111, was named roentgenium after the German discoverer of X-rays, Wilhelm Röntgen. Armbruster won't be drawn on what he has recommended this time, but indicates that patriotism is not the inspiration. 'I did everything to ensure that we do not continue with German scientists and German towns,' he tells me.*

The scene of the most recent element strikes has shifted to Russia. At the Joint Institute for Nuclear Research in Dubna, Yuri Oganessian leads the team that has synthesized elements numbers 114 and 116 (odd-numbered elements are harder to obtain, for reasons to do with nuclear stability). He offers a more personal insight into the quest. 'The work is very difficult, since the probability of a new element nucleus formation is exceedingly small. Very often we get nothing. It may take years,' he says. 'It is not difficult to understand the emotions of the researcher.'

I ask about the distinction between 'finding' and 'making' elements. This excites Oganessian: 'I'd put the question more rudely: why in general do we discover elements?' What, as he puts it, is the need, having synthesized darmstadtium, say, the nineteenth element following uranium, to synthesize a twentieth? Why go on? His answer goes to the heart of what science is about. The discoveries are important less as trophies and more

* In February 2010, IUPAC approved the name copernicium after the astronomer Nicolaus Copernicus, who was born in 1473 in northern Poland, which was then part of the Prussian Confederation.

for what they tell us about the wider world. In Seaborg's heyday, the theoretical model of the atomic nucleus suggested that the catalogue of elements was essentially finite, and that beyond a certain threshold of instability it would actually be impossible to synthesize newcomers. However, advances in theoretical physics during the 1960s then indicated that there might in fact be 'islands of stability' clustered around certain atomic numbers higher up the table. This new understanding has stimulated the hunt for elements that it would have been madness to go after before – and is presumably what encouraged Seaborg to speculate on a periodic table up to atomic number 168. 'Only at the beginning of the present century have we managed to change the method of synthesis and produce elements with atomic numbers from 112 to 118, and prove that the theoretical hypothesis is a reality,' says Oganessian triumphantly.

So are recent discoveries any different from those that have gone before? Oganessian denies it. Each one is a prize in itself, but it also says something about how much further the project can go, perhaps appearing to set a new limit on the number of elements that can exist, or alternatively throwing open new doors of possibility. Its greater significance lies in the contribution it makes to the broader mission of science – the increase of human knowledge. 'Synthesis of a new element is not the end in itself. The efforts of researchers have always been directed to the search for something more important than just filling in the squares of the periodic table. I want to believe that such a motivation doesn't have any exceptions.'

Oganessian and his colleagues have now set their sights on the difficult element number 117. If it turns out to have the properties of a halogen, it will be further proof of the genius of Oganessian's compatriot, Mendeleev. If it doesn't, then it will set chemists wondering anew. 'It looks as if it is going to be one of the most difficult experiments ever carried out.'

# Notes

*Part One: Power*

16 'high degree of sensuous': Veblen, 129.

16 'it is the only': Pliny, 295.

16 'long enough to encircle': Quoted in Chevalier and Gheerbrant, 441.

17 'could be completely': Pliny, 287.

17 'The first person': Pliny, 287.

17 'The second crime': Pliny, 292.

17 'symbol of perversion': Quoted in Chevalier and Gheerbrant, 442.

22 'hard, rebellious quartz': Clemens, 233.

26 'in an entirely uncommercial': Shaw, 8.

26 'the cost of extraction': Herrington, 8.

27 'with production in the best': Herrington, 58.

28 'That gold inwardly taken': Wilkin, 338.

28 'Gold of all the Metals': Geoffroy, 281.

29 'Any fool would know': Quoted in Wilson, 221.

33 'Three months later': Weeks and Leicester, 397.

38 'as shamefully announced': Quoted in McDonald and Hunt, 156.

54 'ochreous stain': Ruskin, 189.

70 'the great burning glass': Knight, 101.

70 'A candle will burn': Faraday, 106. Faraday appears to have made an error in his estimate of the amount of carbonic acid, as 548 tons is equivalent to roughly 1,200,000 pounds.

72 'At some point': Seaborg, 52.

74 'honest-to-God': Seaborg, 72.

74 'Our microchemists isolated': Seaborg, 99.

74 'We briefly considered': Seaborg, 72.

74 'entirely coincidental': Seaborg, 72.

74 'Each element has': Seaborg, 72.

75 'We thought our little': Seaborg, 72.

75 'the UPPU club': Bernstein, 122.

75 'When the husbands': Seaborg, 94.

76 'Plutonium is so unusual': Quoted in Bernstein, 105.

79 'This would be super': Bernstein, 158.

85 'reinforcers': Quoted in Gillispie.

87 'knows everything that happens': Quoted in Gordin, 245.

88 'Symbolic action': Quoted in Gordin, 245.

90 'We thought it fitting': Seaborg, 155.

92 'In mercury the hands': Cocteau in a conversation recorded with André Fraigneau, trs. Vera Traill (London, n.d. [?1952]).

94 'The Chinese have probably': Needham, vol. 13, 143.

100 'As nature produces metals': Roberts, 34.

## Part Two: Fire

118 'Take a Quantity': Quoted in Derham, 187.

119 'Then evaporate all': Quoted in Derham, 187.

124 'curious failure': Vertesi, Janet, 'Light and Enlightenment in Joseph Wright of Derby's "The Alchymist"', Romanticism and the Midlands Enlightenment Conference, Birmingham, UK, 3 July 2004.

125 'a very noble': quoted in William Bray, 244.

127 'The combination': Friedrich, 96.

128 'Dropping phosphorus': Emsley (2000), 158.

131 'the embodiment of the romantic': Haber, 2.

138 'in science, medicine': Chang and Jackson.

144 'through the instrumentality': Quoted in Knight, 97.

155 'ideal of balance': Bensaude-Vincent, 227.

155 'the excessive domination': Bensaude-Vincent, 385.

159 'that insinuating vamp': Harn, 18.

160 'the single most': Lane, 342.

163 'our new metals': Quoted in Quinn, 157.

163 'new emotion': Quinn, 157.

166 'heats the soil': Harvie, 13.

169 'living with a sole': Quoted in Quinn, 156.

176 'the *true* elements': Hartley, 54.

177 'danced about the room': Quoted in Knight, 65.

177 'a most intense light': Quoted in Knight, 66.

177 'the globules often': Quoted in Knight, 66.

178 'the young chemist': Quoted in Hartley, 22.

179 'Do you know where': Quoted in Nechaev, 79.

182 'pitiful and ill conducted': Quoted in Alan Brock, 51.

187 'Have you ever noticed': Quoted in James.

189 'the symptoms of thallium': Sanders and Lovallo, 312–13.

191 'the day of the eclipse': Janssen.

192 'the glorious yellow': Quoted in Weeks and Leicester, 760.

194 'a molecule of gold': Besant and Leadbeater, 1.

194 'arranged on a definite': Besant and Leadbeater, 3.

194 'was seen to consist': Besant and Leadbeater, 9.

195 'so light, and so simple': Besant and Leadbeater, 41.

196 'observed': http://www.chem.yale.edu/~chem125/125/history99/
8Occult/OccultAtoms.html, accessed August 2010.

197 'Manganese offers us': Besant and Leadbeater, 100.

197 'their work would be': Quoted in Nethercot, 52.

## Part Three: Craft

200 'Of the extreme tracts': Herodotus, *Histories*, III.115.

200 'Lusitania': Pliny, *Natural History*, XXXIV; 'brought from the
islands': Pliny, *Natural History*, IV; 'to the north': Strabo, *Geographica*,
III.5.11.

200 'how many historians': Rickard, 339–40.

204 'the respectable material': Levi, 185.

209 'never seen or heard': Levi, 185.

216 'In this way': Freud, 119.

219 'a material for ideas': Quoted in Arasse, 239.

221 'potential to achieve': Auping, 37.

222 'It is, of course': Auping, 39.

228 'for items of luxury': Blair, 213.

237 'the copper covering': Quoted in Jardine, 411.

237 'unashamedly continental': Jardine, 411.

237 'A Ball of Copper': Quoted in Jardine, 317.

238 'large Ball of metall gilt': Quoted in Jardine, 316.

243 'It is the most': Quoted in Thompson, 336.

257 'depends closely': Veblen, 126.

258 'nearly put a shiny': Bryson, 309.

258 'because of its whiteness': Binczewski.

259 'a process of banalization': Hachez-Leroy, 19.

260 'How *aluminium*': Mencken, 415.

262 'lacking only a thatched': Heskett, 171.

263 'turned into Spitfires': Clement, 7.

263 'Spitfires to saucepans': Sparke, 138.

268 'the man who should': Huxley.

273 'one cries out': Vasari, 345.

274 'White was the colour': Festing, 125.

276 'Cleopatra no doubt': Pliny, 137n2.

279 'culturally consequent': Sennett, 144; 'The attribution of ethical':
Sennett, 137.

281 'soft, buttery look': Quoted in *American Scientist*, July–August 1998.

282 'they look more like': Storrie, 166.

## Part Four: Beauty

289 'at the best is': Quoted in Feller, vol. 1, 71.

294 'should not have': Elliott et al.

297 'rolling sculpture': Drexler, foreword.

298 'from wine jugs': Bayer, Gropius and Gropius, 134.

304 'This relationship of woman': Quoted on the Tate website: www.tate.org.uk.

304 'no metal, it seems': Quoted in advertisement copy, *New Yorker*, 28 October 1933, 36.

304 'there is little wrong': Keats.

306 'precious sheets of sapphire': Quoted in Gage, 72.

307 'luminous darkness': Gage, 71.

319 'doing great service': MacCarthy, 351.

319 'a very dangerous piece': Meharg, 688; 'Killer Wallpaper', *Spectroscopy Europe*, 16 (5) (2004), 16–19.

324 'Such a request': 'Much Ado about Vanadium', *Ascidian News*, 51 (June 2002), at http://depts.washington.edu/ascidian/AN51.html.

325 'Long ago there lived': Quoted in Jorpes, 74.

326 'In his naming': Jorpes, 75.

327 'the most esteemed': Guyton de Morveau et al., vol. 6, 64; vol. 4, 231.

332 'a most astonishingly': Ramsay and Rayleigh, 187.

332 'Drink to the gas': Quoted in Travers (1956), 141.

333 'You get a new element': Quoted in Travers (1956), 174.

334 'As Ramsay pressed': Travers (1956), 178.

341 'marriage chapels': Venturi, Scott Brown and Izenour, 20.

342 'caught some very good': Boyd, 157.

344 'an apparatus for chymical': Read, 130.

348 'the Solar Calcination': Read, 101–7.

## Part Five: Earth

360 'an enormous undertaking': C. H. Evans, xviii.
361 'any further investigation': Suyver and Meijerink.
374 'It is not without great': Quoted in C. H. Evans, 8.

## Epilogue

391 'for some moments': Travers (1956), 178.
393 'incredibly touched': Seaborg, 254.

# References and Select Bibliography

Agricola, Georgius, *De Re Metallica*, trs. Herbert Clark Hoover and Lou Henry Hoover (New York: Dover, 1950)

Arasse, Daniel, *Anselm Kiefer* (London: Thames and Hudson, 2001)

Armstrong, Lyn, *Woodcolliers and Charcoal Burning* (Horsham, Sussex: Coach Publishing House, 1978)

Auping, M., ed., *Anselm Kiefer: Heaven and Earth* (London: Prestel, 2005)

Ball, Philip, *Bright Earth: The Invention of Colour* (London: Penguin, 2001)

Ball, Philip, *$H_2O$: A Biography of Water* (London: Weidenfeld and Nicolson, 1999)

Ball, Philip, *The Ingredients* (Oxford: Oxford University Press, 2002)

Batchen, Geoffrey, *Burning with Desire: The Conception of Photography* (Cambridge, MA: MIT Press, 1997)

Bayer, Herbert, Walter Gropius and Ise Gropius, eds., *Bauhaus 1919–1928* (London: Secker and Warburg, 1975)

Bayfield, Gerald, *Dereham's Forgotten Scientist William Hyde Wollaston* (Dereham, Norfolk: Dereham Antiquarian Society, 1990)

Belcher, Captain Sir Edward, *Narrative of a Voyage Round the World, performed in Her Majesty's Ship Sulphur, during the years 1836–1842* (London: Henry Colburn, 1843)

Bensaude-Vincent, Bernadette, *Lavoisier: Mémoires d'une révolution* (Paris: Flammarion, 1993)

Bernstein, Jeremy, *Plutonium: A History of the World's Most Dangerous Element* (Washington, DC: Joseph Henry Press, 2007)

Besant, Annie, and C. W. Leadbeater, *Occult Chemistry* (London: Theosophical Publishing House, 1919)

Binczewski, George J., 'The Point of a Monument: A History of the Aluminum Cap of the Washington Monument', *JOM*, 47 (11) (1995), 20–25

Blair, Claude, ed., *The History of Silver* (London: Macdonald, 1987)

Bostock, John, *An Elementary System of Physiology* (London: Baldwin, Cradock and Joy, 1824–7)

Boyd, Brian, *Vladimir Nabokov: The American Years* (Princeton: Princeton University Press, 1993)

Bray, Warwick, *The Gold of El Dorado* (London: Times Books, 1978)

Bray, William, ed., *The Diary of John Evelyn*, vol. 2 (New York: M. Walter Dunne, 1901)

Brock, Alan St H., *A History of Fireworks* (London: Harrap, 1949)

Brock, William H., *The Fontana History of Chemistry* (London: Fontana, 1992)

Bryson, Bill, *A Short History of Nearly Everything* (London: Doubleday, 2003)

Cameron, A. D., *Tarnished Silver* (New York: Midmarch Arts, 1996)

Chang, Hasok, and Catherine Jackson, eds., *An Element of Controversy: The Life of Chlorine in Science, Technology, Medicine and War* (London: British Society for the History of Science, 2007)

Chevalier, Jean, and Alain Gheerbrant, *The Penguin Dictionary of Symbols*, trs. John Buchanan-Brown (London: Penguin, 1996)

Clark, Grahame, *Symbols of Excellence: Precious Materials as Expressions of Status* (Cambridge: Cambridge University Press, 1986)

Clemens, Samuel, *The Writings of Mark Twain*, vol. 7 (Hartford, CT: American Publishing Company, 1901)

Clement, Mark, *Aluminium: A Menace to Health* (London: Faber and Faber, 1941)

Cologni, Franco, and Eric Nussbaum, *Cartier le joaillier du platine* (Paris: Bibliothèque des Arts, 1995)

Conrad, Peter, *Modern Times, Modern Places* (London: Thames and Hudson, 1998)

Cotterell, Arthur, *Norse Mythology* (New York: Anness Publishing, 2000)

Craddock, Paul T., *Early Metal Mining and Production* (Edinburgh: Edinburgh University Press, 1995)

Crossley-Holland, Kevin, *The Penguin Book of Norse Myths* (London: Penguin, 1993)

Daintith, John, and Derek Gjertsen, *A Dictionary of Scientists* (Oxford: Oxford University Press, 1999)

Davis, Donald W., and Randall A. Detro, *Fire and Brimstone: The History of Melting Louisiana's Sulphur* (Baton Rouge: Louisiana Geological Survey, 1992)

Derham, W., *Philosophical Experiments and Observations of the Late Eminent Dr Robert Hooke FRS . . . and Other Eminent Virtuoso's of his Time* (London: Innys, 1726)

Donovan, Arthur, *Antoine Lavoisier: Science, Administration and Revolution* (Cambridge: Cambridge University Press, 1993)

Drakard, David, and Paul Holdway, *Spode Transfer Printed Ware 1784–1833* (Woodbridge, Suffolk: Antique Collectors' Club, 2002)

Drexler, Arthur, *Eight Automobiles* (New York: Museum of Modern Art, 1951)

Eliade, Mircea, *The Forge and the Crucible* (London: Rider, 1962)

Elliott, P. J., C. J. C. Phillips, B. Clayton and P. J. Lachmann, 'The Risk to the United Kingdom Population of Zinc Cadmium Sulfide Dispersion by the Ministry of Defence during the "Cold War"', *Occupational and Environmental Medicine*, 59 (2002), 13–17.

Emsley, John, *Nature's Building Blocks: An A–Z Guide to the Elements* (Oxford: Oxford University Press, 2001)

Emsley, John, *The Shocking History of Phosphorus: A Biography of the Devil's Element* (London: Macmillan, 2000)

Emsley, John, *Vanity, Vitality, and Virility: The Science Behind the Products You Love to Buy* (Oxford: Oxford University Press, 2004)

Evans, B. Ifor, *Literature and Science* (London: George Allen and Unwin, 1954)

Evans, C. H., ed., *Episodes from the History of the Rare Earth Elements* (Dordrecht: Kluwer, 1996)

Faraday, Michael, *The Chemical History of a Candle* (London: Chatto and Windus, 1908)

Feller, Robert L., ed., *Artists' Pigments: A Handbook of Their History and Characteristics*, vol. 1 (Cambridge: Cambridge University Press, 1986)

Festing, Sally, *Barbara Hepworth: A Life of Forms* (London: Viking, 1995)

Fowles, G., *Lecture Experiments in Chemistry* (London: G. Bell and Sons, 1963)

Freud, Sigmund, 'The Theme of the Three Caskets', in *Writings on Art and Literature* (Stanford, CA: Stanford University Press, 1997).

Friedrich, Jörg, *The Fire: The Bombing of Germany, 1940–1945*, trs. Alison Brown (New York: Columbia University Press, 2006)

Gage, John, *Colour and Culture* (London: Thames and Hudson, 1993)

Geoffroy, E.-F., *A Treatise of the Fossil, Vegetable and Animal Substances That Are Made Use of in Physik* (London: Innys, 1736)

Gillespie, C. C., ed., *Dictionary of Scientific Biography* (New York: Scribner's, 1974)

Gordin, Michael D., *A Well-Ordered Thing: Dmitrii Mendeleev and the Shadow of the Periodic Table* (New York: Basic, 2004)

Greenberg, Arthur, *The Art of Chemistry* (Hoboken, NJ: Wiley, 2003)

Gribbin, John, *Science: A History* (London: Allen Lane, 2002)

Grissom, Carol, *Zinc Sculpture in America: 1850 to 1950* (Newark, NJ: University of Delaware Press, 2009)

Guyton de Morveau, L.-B., et al., *Encyclopédie méthodique chimie, pharmacie et metallurgie* (Paris: Panckoucke, 1786–1815)

Haber, L. F., *The Poisonous Cloud: Chemical Warfare in the First World War* (Oxford: Clarendon Press, 1986)

Hachez-Leroy, Florence, *L'Aluminium français* (Paris: CNRS Editions, 1999)

Hampel, Clifford A., ed., *The Encyclopedia of the Chemical Elements* (New York: Reinhold, 1968)

Harn, Orlando C., *Lead, The Precious Metal* (London: Jonathan Cape, 1924)

Hartley, Harold, *Humphry Davy* (London: Nelson, 1966)

Harvie, David I., *Deadly Sunshine: The History and Fatal Legacy of Radium* (Stroud, Glos.: Tempus, 2005)

Haynes, William, *The Stone that Burns* (New York: Van Nostrand, 1942)

Hearn, Chester G., *Circuits in the Sea* (Westport, CT: Praeger, 2004)

Henderson, Julian, *The Science and Archaeology of Materials* (London: Routledge, 2000)

Herrington, Richard, Chris Stanley and Robert Symes, *Gold* (London: Natural History Museum, 1999)

Heskett, John, *Industrial Design* (London: Thames and Hudson, 1980)

Hirsch, Robert, *Seizing the Light: A History of Photography* (New York: McGraw-Hill, 2000)

Hurlbut, Cornelius S., Jr, and Robert C. Kammerling, *Gemology*, 2nd edn (New York: Wiley, 1991)

Hutchinson, John, et al., *Antony Gormley* (London: Phaidon, 2001)

Huxley, Thomas, 'On a Piece of Chalk', *Macmillan's Magazine*, 18 (1868), 396–408

James, Frank A. J. L., 'Of "Medals and Muddles" the Context of the Discovery of Thallium: William Crookes's Early Spectro-Chemical Work', *Notes and Records of the Royal Society*, 39 (1984), 91–104

Janssen, M., 'The Total Solar Eclipse of August 1868', *Astronomical Register*, 7 (1869), 107–10

Jardine, Lisa, *On a Grander Scale: The Outstanding Career of Sir Christopher Wren* (London: HarperCollins, 2002)

Jorpes, J. Erik, *Jac. Berzelius: His Life and Work*, trs. Barbara Steele (Berkeley: University of California Press, 1970)

Keats, John, *The Insolent Chariots* (Philadelphia: Lippincott, 1958)

Knight, David, *Humphry Davy: Science and Power* (Oxford: Blackwell, 1992)

Lamont-Brown, Raymond, *Humphry Davy: Life Beyond the Lamp* (Stroud, Glos.: Sutton, 2004)

Lane, Nick, *Oxygen: The Molecule that Made the World* (Oxford: Oxford University Press, 2002)

Lecoq de Boisbaudran, P.-E., *Spectres lumineux* (Paris: Gauthier-Villars, 1874)

Levi, Primo, *The Periodic Table* (London: Michael Joseph, 1985)

Lister, T., *Classic Chemistry Demonstrations* (London: Royal Society of Chemistry, 1995)

Loring, F. H., *The Chemical Elements* (London: Methuen, 1923)

MacCarthy, Fiona, *William Morris* (London: Faber and Faber, 1994)

McDonald, Donald, and Leslie B. Hunt, *A History of Platinum and Its Allied Metals* (London: Johnson Matthey, 1982)

McEwan, Colin, ed., *Pre-Columbian Gold: Technology, Style and Iconography* (London: British Museum Press, 2000)

McGee, Harold, *On Food and Cooking: The Science and Lore of the Kitchen*, 3rd edn (London: HarperCollins, 1991)

McLynn, Frank, *Napoleon: A Biography* (London: Pimlico, 1998)

Man, John, *The Terracotta Army* (London: Bantam, 2007)

Meharg, Andrew A., 'Science in Culture', *Nature*, 423 (2003), 688

Meharg, Andrew A., *Venomous Earth* (Basingstoke: Macmillan, 2006)

Melhado, Evan M., and Tore Frängsmyr, eds., *Enlightenment Science in the Romantic Era: The Chemistry of Berzelius and Its Cultural Setting* (Cambridge: Cambridge University Press, 1992)

Mencken, H. L., *The American Language* (New York: Knopf, 1955)

Mèredieu, Florence de, *Histoire matérielle et immatérielle de l'art moderne* (Paris: Bordas, 1994)

Morris, Richard, *The Last Sorcerers: The Path from Alchemy to the Periodic Table* (Washington, DC: Joseph Henry Press, 2003)

Nassau, Kurt, *The Physics and Chemistry of Color* (New York: Wiley, 2001)

Nechaev, I., *Chemical Elements* (London: Baker and Walls, 1944)

Needham, Joseph, *Science and Civilisation in China* (Cambridge: Cambridge University Press, 1954–2008)

Nethercot, Arthur H., *The Last Four Lives of Annie Besant* (Chicago: University of Chicago Press, 1963)

Newton Friend, John A., *Man and the Chemical Elements* (Newark, NJ: Charles E. Graham, 1951)

Pastoureau, Michel, *Blue: The History of a Color* (Princeton: Princeton University Press, 2001)

Pearce, Emma, *Artists' Materials* (London: Arcturus, 2005)

Perkowitz, Sidney, *Empire of Light* (Washington, DC: Joseph Henry Press, 1996)

Pliny the Elder, *Natural History: A Selection* (London: Penguin, 2004)

Quinn, Susan, *Marie Curie: A Life* (London: Heinemann, 1995)

Ramsay, William, and Lord Rayleigh, 'VI. Argon: A New Constituent of the Atmosphere', *Philosophical Transactions of the Royal Society*, 186A (1895), 187

Read, John, *Humour and Humanism in Chemistry* (London: G. Bell and Sons, 1947)

Rhodes, Richard, *The Making of the Atomic Bomb* (New York: Simon and Schuster, 1986)

Rickard, T. A., *Man and Metals* (New York: McGraw-Hill, 1932)

Roberts, Gareth, *The Mirror of Alchemy: Alchemical Ideas and Images in Manuscripts and Books from Antiquity to the Seventeenth Century* (London: British Library, 1994)

Roy, Ashok, ed., *Artists' Pigments: A Handbook of Their History and Characteristics*, vol. 2 (Washington, DC: National Gallery of Art, 1993)

Ruskin, John, *The Two Paths* (London: Smith Elder, 1859)

Sacks, Oliver, *Uncle Tungsten* (London: Pan Macmillan, 2001)

Sanders, Dennis, and Len Lovallo, *The Agatha Christie Companion* (London: W. H. Allen, 1985)

Scerri, Eric R., *The Periodic Table: Its Story and Its Significance* (Oxford: Oxford University Press, 2007)

Schama, Simon, *Landscape and Memory* (London: Harper Perennial, 1995)

Seaborg, Glenn, *Adventures in the Atomic Age* (New York: Farrar, Straus and Giroux, 2001)

Sebald, W. G., *The Rings of Saturn* (London: Harvill, 1998)

Seibel, Clifford W., *Helium: Child of the Sun* (Lawrence, KS: University Press of Kansas, 1968)

Sennett, Richard, *The Craftsman* (London: Penguin, 2008)

Shaw, Bernard, *The Perfect Wagnerite: A Commentary on the Niblung's Ring* (New York: Brentano's, 1916)

Sinkankas, John, and Peter G. Read, *Beryl* (London: Butterworth, 1986)

Sparke, Penny, *An Introduction to Design and Culture in the Twentieth Century* (London: Allen and Unwin, 1986)

Storrie, Calum, *The Delirious Museum* (London: I. B. Tauris, 2006)

Strathern, Paul, *Mendeleyev's Dream: The Quest for the Elements* (London: Hamish Hamilton, 2000)

Suyver, F., and A. Meijerink, 'Europium beveiligt de Euro', *Chemisch2Weekblad*, 98, 4 (2002), 12–13

Szydlo, Andrew, *Water Which Does Not Wet Hands: The Alchemy of Michael Sendivogius* (Warsaw: Polish Academy of Sciences, 1994)

Taylor, Sherwood F., *The Alchemists: Founders of Modern Chemistry* (London: Heinemann, 1951)

Thompson, Silvanus P., *The Life of William Thomson, Baron Kelvin of Largs* (London: Macmillan, 1910)

Travers, Morris W., *Sir William Ramsay* (London: Edward Arnold, 1956)

Travers, Morris W., *The Discovery of the Rare Gases* (London: Edward Arnold, 1928)

Trifonov, D. N., and V. D. Trifonov, *Chemical Elements: How They Were Discovered* (Moscow: Mir, 1982)

Tylecote, R. F., *A History of Metallurgy*, 2nd edn (London: Institute of Materials, 1992)

Vasari, Giorgio, *The Lives of the Artists*, vol. 1 (London: Penguin, 1987)

Veblen, Thorstein, *The Theory of the Leisure Class* (Amherst, NY: Prometheus, 1998)

Venturi, Robert, Denise Scott Brown and Steven Izenour, *Learning from Las Vegas* (Cambridge, MA: MIT Press, 1977)

Wagner, Monika, *Das Material der Kunst* (Munich: C. H. Beck, 2001)

Webster Smith, B., *Sixty Centuries of Copper* (London: Hutchinson, 1965)

Weeks, Mary E., and Henry M. Leicester, *Discovery of the Elements*, 7th edn (Easton, PA: Journal of Chemical Education, 1968)

White, Michael, *Isaac Newton: The Last Sorcerer* (London: Fourth Estate, 1997)

Wilkin, Simon, ed., *Sir Thomas Browne's Works* (London: William Pickering, 1835)

Wilkinson, J. B., and R. J. Moore, eds., *Harry's Cosmeticology*, 7th edn (London: George Godwin, 1982)

Wilson, Arthur, *The Living Rock* (Abington, Cambs.: Woodhead Publishing, 1994)

Zelizer, Barbie, *Visual Culture and the Holocaust* (Piscataway, NJ: Rutgers University Press, 2000)

# Text Credits

Every effort has been made to trace copyright holders and to obtain their permission for the use of copyright material. The publisher apologizes for any errors or omissions and would be grateful to be notified of any correction that should be incorporated in future editions of this book.

Extract from *The Waste Land* © the Estate of T. S. Eliot by permission of Faber and Faber Ltd.

Extract from 'Cargoes' by John Masefield reproduced by permission of The Society of Authors as the Literary Representative of the Estate of John Masefield.

Extract from *Dr. Strangelove*, copyright © 1963, renewed 1991, Columbia Pictures Industries, Inc. All rights reserved. Courtesy of Columbia Pictures.

Extract from *A Streetcar Named Desire* by Tennessee Williams, published by Penguin. Copyright © The University of the South, 1947; Copyright renewed © The University of the South, 1975. Reproduced by permission of Georges Borchardt Inc. in the UK. Reproduced by permission of New Directions Publishing Corp. in the USA.

Entnommen Übersetzungen aus Zuck, poem Bertolt Brecht, 1976
Eine fremdsprache Bühnen und Funkrecht sowie Abdruck und Eye
Suhrkamp Verlag Frankfurt am Main 1976

Translation from "Poems Home to Gerong" by W. Brown etc.
Publication law the Estate of R. Buckminster Fuller.

# Index

The elements of the modern periodic table are indicated in **bold**.

Read on for a preview of

*Tide: The science and lore of the
greatest force on earth*

Coming soon from

Hugh Aldersey-Williams

# Introduction

Halfway along the causeway that crosses the sandflats and marshes to the holy Northumbrian island of Lindisfarne, with its ancient priory and castle, is a little refuge, a hut built on stilts with wooden steps up to the door. Notices fixed to the door state the purpose of this odd structure. It is there for the use of unwary travellers who find themselves caught out on the three-mile crossing by the rising tide. Among the notices is a sun-bleached cartoon from a local newspaper. It shows a family huddled together on the roof of their car as the rising seawater begins to lap at its windows. The caption has one parent complaining to the other: 'I didn't think that tides applied to tourists.'

On the islands of Britain, we live surrounded by water that rises and falls according to rules that are a mystery to almost all of us. We are adrift on a sea whose movements we do not understand. Many are the summers on the stretch of coast where I live in Norfolk that a child is swept out to sea on the tide, to be washed up dead days or months later many miles along the coast. Half of the world's population lives in coastal regions lapped by tidal waters, and yet, unless we depend directly upon the ocean for our living, we remain inattentive to its strange rhythms and ignorant of their complicated causes.

We do not understand, therefore, why the pleasure boat trip cannot depart on the Sunday morning that would fall in best with our weekend plans. We do not understand why the sandy beach we remember from a visit long ago is today a strip of uninviting shingle. We do not understand that the reason why we live the way we do, why we speak the language that we do, may be traceable to a particular seaborne invasion or a particular naval battle whose outcome hinged on the turning of a single tide. We do not understand why the sessile oyster, which depends on each new tide to replenish its food supply, was once a staple food of the destitute. We do not understand our dependence on the astonishing diversity of life that inhabits the intertidal zone, or appreciate that the evolution of life itself may have been reliant upon the ebb and flow of great oceans to create the conditions under which the correct chemical ingredients could mingle to produce the first primitive organisms.

The tide? What's that got to do with the price of fish? Well, a very great deal as it happens. The American variant of this expression is: *What's that got to do with the price of tea in China?* Well, the tide may not much affect the price of tea in China, but it has been important in determining the price of Chinese tea, for in the days when tea clippers raced round the Cape of Good Hope to be first to bring the new-picked leaves to London or Boston, the state of the tide as they closed upon their port of destination could decide the winner, who was then able to sell his cargo at a premium. The tide has always influenced where important marine resources are concentrated; and over the centuries, the tide has often been a hidden factor in determining the location of great ports and in governing their subsequent success in commerce.

My own formative encounter with the tide happened in waters just off the coast of the Isle of Wight, a locale of such comical tameness, lapping an ice-cream island of garden gnomes and miniature railways, that it still amazes me years later that we could have been in mortal danger.

We were sailing from Weymouth to Yarmouth in the wooden boat that my father had hand-built over a period of twelve years. Over the ground, the distance was thirty miles or so, which would be a full day's sailing in our small craft. The journey would take us eastward along the Dorset coast past the Isle of Purbeck and Bournemouth and then up the Needles channel into the Solent. The Needles channel was a tricky stretch of water, bounded by a gravel bank known as the Shingles on one side and by the jagged promontories of the Isle of Wight on the other. The tide runs fast, especially at Hurst Point where a spit juts out from the mainland, forcing the waters through a gap less than a mile wide.

It is also one of the world's busiest shipping lanes, and for this reason has been generously strewn with navigation buoys – red ones to denote the port side of the channel, green ones for starboard – and others known as cardinal marks that indicate the location of isolated underwater obstructions. None of these markers in the least resemble the squashy plastic buoys you see used for moorings in sheltered waters. They are placed by Trinity House, the national authority responsible for safe 'signposting' at sea, and are substantial objects to find floating in the water. The cardinal marks in particular are imposing steel structures, tall gantries rising to the height of a double-decker bus above the waves, with just as much ironwork hidden below the waterline in buoyancy compartments and ballast. Some three metres in diameter and weighing six tonnes, they are not the kind of thing you argue with in a vessel of any size, and certainly not in a small wooden yacht.

Any passage under sail must be planned bearing in mind the likely weather conditions, in particular the strength and direction of the wind, and also the tidal movements likely to be encountered during the expected period at sea. A journey of a few hours under favourable conditions might not be possible at all if attempted at the wrong state of the tide. In the English Channel, as in most coastal bodies of water around the world, the tide runs for approximately six hours in one direction before reversing and running in the opposite direction for the next six hours. Our thirty-mile route was long enough that we would not be able to complete the whole journey with the tide helping us all the way.

The navigational challenge, then, was to decide upon a time of departure that would ensure we had the tide running with us for as much of the journey as possible, and especially when crossing those stretches of water where the tidal stream ran fastest, where we might otherwise be brought to a halt or even find ourselves going backwards in relation to the land. In particular, we had to time our trip to avoid the 'tidal gate' being against us at Hurst Point where the current can run at four knots or more – faster than we could sail under most conditions and faster than we could make way under engine in even the calmest sea.

We had all the information we needed. We had the right charts – *Bill of Portland to The Needles* and *Solent: Western Approaches* – which I remember my mother kept religiously up to date using a special violet ink sold for the express purpose of making the official chart corrections issued in weekly Admiralty Notices to Mariners. We also had a tidal atlas, a booklet with pages showing the same body of water at each hour before and after high water for a full tidal cycle of twelve hours. By means of arrows of varying thickness, each map denoted the expected speed and direction of the tide at the given hour across the entire charted area of sea.

My father made his calculation, and we cast off from the quay at Weymouth that September morning at ten past eight on the falling tide. The aim was to sail against the tide where it was weak in Weymouth Bay, and then on through slack water before using the rising tide to carry us into the Solent. By the time we reached the gate, all being well, we would have a gently favourable tide running with us and would be home in Yarmouth before dark.

The wind was slow to get up that morning, and we were obliged to motor for several hours before a gentle breeze filled in from the west, allowing us to set the spinnaker. We made soporific progress eastward until mid-afternoon when the outline of the Isle of Wight finally picked itself out of the haze.

Some hours later, in now fading light, we passed the Needles. The tide, which had been running gently against us, had by now turned in our favour. We peered ahead for the next marks: Shingle Elbow, which was due to appear ahead of us to port, and its opposite number marking the starboard side of the channel, a buoy named Bridge. This latter was a cardinal mark placed at the western extremity of an underwater ridge of rocks extending out from the Needles (and extending, in geological terms, as a band under the sea to the chalk of Purbeck that we had left behind earlier in the day). If we could have seen them, the cones on top of the mark would have been pointing

towards each other, telling us that we should steer to the
west of it. Instead, it being now nearly dark, we identified
the buoy by its light, marked on the chart as VQ (9)10s,
meaning that it emits nine very quick flashes every ten seconds. It was hard to gauge its distance, but it seemed to be
some way off on our starboard bow.

We sailed slowly on through the syrupy water, the light
of the buoy getting closer with each strobe-like sequence
of flashes. These flashes, though, did not begin to slip past
us on our starboard side as we expected with a tide bearing
us straight up the channel, but remained on a constant
bearing to our heading. With each ten-second burst, the
lights grew brighter and larger, but stayed stubbornly forward of us at the same angle off our bow. Slowly, we began
to understand what was happening. Our boat was sliding
sideways as well as sailing forwards. The tide, which we
had assumed to be running along the channel, was in fact
setting strongly across it, sweeping us eastwards, closer to
the submerged ridge. Our deliberate motion through the
water combined with our drifting on the tide was holding
us on a collision course with the buoy. In fact, our speed
through the water was modest. It seemed it was the buoy
that was making the greater headway, as we now became
horribly aware, hearing the sloshing of the 'bow wave' it
was making as it apparently forged through the tidal current towards us.

From the land, I had often observed the Solent tides
ripping past the buoys set close to the island shore, the
buoys leaning with the strain on the cables anchoring them
to the seabed, shouldering the force of the current, and
streaming out a foamy wake as if they were ferries running
to a schedule. I had occasionally drifted off to sleep at

home on a stormy night hearing the distant wave-made sound of the clanging bell of one of these buoys, even though it was fully a mile away. Inexorably, every six hours or so, the tide would turn and each buoy would shuttle back on its endless tethered commute. In between times, near high or low water when the tide was slack, the metal buoy would lie massive and silent, as if nothing on earth could ever be so powerful as to give it this impression of urgent business. Doubtless, in an hour or two, the Bridge buoy would come to such a rest. But this was no help to us now.

To counter the effect of the tide, we swiftly altered course, my father pushing the tiller away as the rest of us trimmed the sails in the hope that we could point closer to the wind and overhaul the buoy. The manoeuvre made no difference. The combined angle of our sailing direction and the flow of the tide still held our fragile wooden boat on course to strike six tonnes of Trinity House steel. The next burst of 'very quick' flashes took on an edge of urgency. Suddenly, I saw that we were not going to be able to get past the buoy on the correct side – the boat would simply stall if we headed further up into the wind and we would be powerless to do anything. Impulsively, I seized the tiller and pulled it sharply to windward, so that the boat bore away from the wind, and we shot past the buoy on the down-tide side – the 'wrong' side – seemingly just inches away.

Out of the pompous belief that it was the proper thing to do aboard any ship, we had always kept a logbook. In it, one of us would note the details of even the shortest sailing excursion at the end of the day – the weather conditions, the times and places of departure and arrival, and so on. Occasionally, there were instructive remarks about the speed achieved in a given wind with this sail as opposed to that, or the particular arrangements for mooring in an unfamiliar harbour. Quite a few measurements we recorded were concerned with maximizing the speed we could reach under engine, noting variations in the pitch of the propeller and the revolutions of the engine that would nudge it ever closer to a speed of four knots, which seemed to be a theoretical maximum limit. These experiments reflected our acute consciousness that the tide in these waters could at times run at least this fast, leaving us potentially at its mercy.

Yet the log reveals little of the drama of that September night. The day's recorded run was 52.77 nautical miles, almost twice the actual distance over the seabed, showing that much of our journey had been a slow battle against the tide, and that our progress overall had been hindered rather more than it had been helped. 'Now completely dark,' reads the entry for 20.00 hours. 'Very nearly hit Bridge Buoy due to set of tide while steering by Hurst leading lights.' The subtext is clear enough. We did not nearly hit the buoy because of our own inexperience or poor seamanship, thank you. We were doing the right thing, steering by the lights. No. We nearly hit the buoy 'due to set of tide'. The tide was to blame. The tide, that capricious, malevolent, world-dominating force.

This is not a book about the sea. It does not feature long days before the mast, scurvy, whaleboats, pirates, ship's biscuits and tots of rum. It does not brave hurricanes or typhoons. It does not centre on a man condemned for ever to sail the oceans, nor does it have mermaids to lure him astray. It is not about the vasty deep, and man's battle with it, and his often curious reasons for going into that battle in the first place.

This is a book about the sea. It is about the sea that we all know – the beach, the coast, land's edge, land's end. The sea we regard as our holiday playground, yet understand hardly at all. The sea that we move cautiously upon, and that moves us in mysterious ways, both physically and emotionally.

The tides are complicated and some people find they obtain their most elegant explanation in the form of mathematics. Yet even if you are able to interpret the symbols and equations, you immediately lose the visceral sense of the oceans' rising and falling, the sense of the tide's upper hand over humankind's maritime adventuring. The tide has simple physical power over small craft at sea, but it also has power over the senses and the mind. To stand and gaze out from aboard a boat held at anchor upon a tidal body of water within sight of land is to subject oneself to a hallucination of bewildering intensity. For during those regular periods when the tide is running strongly in one direction or another, it will seem that the boat is cleaving purposefully through the sea even though reason tells you it is going nowhere. It is a powerful and disturbing illusion, enhanced in its hypnotic effect perhaps by sunbeams bouncing off the streaming water or the rhythmic undulation of the boat, or by your own hunger or thirst. You blink hard to dismiss it, but when you open your eyes it is still there – the boat is definitely moving forwards, anchor chain and all, this last determinedly probing the waters ahead like a narwhal's tusk.

After a while, your brain reframes the illusion. Mesmerized by the scintillating wavelets, it now conceives that it is the water that is static while you and the boat are racing together with the distant land towards some shared destination. The surface of the water may be glassy and smooth, wrinkled only slightly by the deep turbulence of the tidal flow itself, or it may be choppy, as it is when the wind blows against the tide and friction between the air and the water kicks up short, steep waves. It doesn't matter. Your impression nevertheless is that all this water must be essentially stationary. For what great power could convey such a mass of ocean back and forth so swiftly without apparent effort or cause?

We learn at school, of course, that the main answer to this question is the moon and its gravitational pull on the earth, and by adulthood we have assimilated this information without demur. We stupidly take the force of gravity as read, and reserve our wonder for more modern oddities such as quantum theory or dark matter. Yet how very odd the colossal, invisible force of gravity still is if we stop to think about it at all. With the rushing tide, we have a visible, soaking, undeniable expression of that weirdness.

Before we understood them in a scientific sense, the tides were already comprehended in their way by myth makers and storytellers. The actual power of the tide to drag sailors to their death is surely enough to explain the lure of the siren and the sucking tentacles of the kraken. The sea has no need of assistance from malevolent creatures, which are the mere invention of ignorant seafarers, made up because they offer a more believable story than the horrid idea that something as routine and automatic as the unaided tide has the power to take a person's life. The fabulous sea monsters that adorn the peripheries of old explorers' maps may similarly be unfamiliar deep-sea species whirled up into the light by tidal upwellings, while the tidal bores of various rivers earn themselves the names of gods and monsters.

Scientific knowledge offers an alternative explanation for some of these stories, but the stories live on. Science has begun to investigate the mysteries of the oceans – only just begun, really: oceanography is one of the youngest of the sciences, and the subject is all too obviously a large one. The investigation of the oceans has begun, but we can hardly be said to have tamed them. We can now predict the tide theoretically to a high degree of accuracy – far more accurately, in fact, than it ever occurs in our real experience, where other factors interfere with it. Why the drive to know in such obsessive detail? In part, it is because the tides provide a perfect example of the complex problem that should be exactly soluble. All the variables are known, it is only a matter of doing the sums. The tide offers an irresistible mathematical tease, which is undoubtedly why it has historically attracted some of the world's finest physicists and astronomers. But there are also practical reasons why it is important to have precise answers beyond the needs of navigators and fishermen, and these turn out to have relevance for the future of us all.

There is a good reason why, despite all this, there has been no accessible book about the science of the tides. It is because the topic swiftly becomes more complex than lends itself to explanation in words. The author of a field guide to the salt marshes of New England gave his drily brief synopsis of the subject this heading: 'Tides in Perhaps More Detail Than I Should Include in This Book'. I know how he feels. Harmonic equations give the scientist a far more versatile tool for understanding what goes on.

But I know that I cannot follow this route – for your sake, and for mine.

Instead, I have tried a different approach. I have given an episodic history of the science of tides from the earliest times up until the present day. This course allows me to voyage from the earliest science of Aristotle, who is said to have drowned himself when he failed to figure out the Greek tides, to the better informed investigations of Galileo and Newton, and then on to a scientific understanding so complete that we are now able to predict the tides with that high degree of accuracy (far greater than any sailor needs), which is yielding important new evidence of our own impact on this watery planet. Along the way, we drop anchor in less familiar harbours, too, pausing to acknowledge the unexpected contribution to the understanding of the tides by figures such as Bede and St Thomas Aquinas, men not primarily thought of as scientists, but whose great minds could not ignore this cosmic puzzle.

I have interwoven the scientific strand of my narrative with two others: one in which stories of events – be they historical, artistic or entirely fabulous – where the tide plays a crucial role get their due; and another in which I go in search of special places made by the tide. I include these episodes to show that the tide is not only a scientific challenge, but also a force of a quite different kind – a physical and a psychological influence on our culture whose presence cannot be denied. The tides have determined the course of battles and have inspired poets and artists. And they continue to do so today.

In weaving these strands together, strict chronology is occasionally sacrificed for the sake of a thematic connection. The science is, I hope, made simple but not simplistic.

This is not a textbook about the tides. It is a book of stories and journeys. I hope you find yourself able to go with the flow. It is always unwise to fight the tide.